CONTENTS

KW-222-107

Waste Strategy 2000

for England and Wales

Part 2

Presented to Parliament by the Secretary of State for
the Environment, Transport and the Regions
by Command of Her Majesty:
Laid before the National Assembly for Wales by the First Secretary:
May 2000

Cm 4693-2

£20

Department of the Environment, Transport and the Regions
Eland House
Bressenden Place
London SW1E 5DU
Telephone 020 7944 3000
Internet service http://www.detr.gov.uk

ISBN 0 10 146933 0

Printed in the UK on material containing 75% post-consumer waste and 25% ECF pulp.

May 2000.

CHAPTER 1

Introduction

1.1   Part 1 of this strategy sets out a vision of sustainable waste management in England and Wales for the next 20 years. It offers a strategic overview of waste policy, outlines the scale of the task facing us and the tools we can bring to bear on that challenge, and gives details of the actions stakeholders need to take in the next 5 years to meet the vision and targets we have set ourselves.

1.2   Part 2 is set out as a complement to Part 1, and should be read in conjunction with that document. In it, we:

- provide further data on the nature and quantity of waste production

- provide more detailed background to many of the policies described in Part 1

- describe some of the progress we have made since the last waste strategy, *Making Waste Work*, was published in 1995

- set out arrangements for a number of specific waste streams, including packaging waste and special (hazardous) waste

- describe the existing facilities for managing waste in England and Wales

The vision, aims and objectives of the waste strategy

1.3   The key messages of the waste strategy are:

- we produced 106 million tonnes of commercial, industrial and municipal waste in England and Wales last year, most of which was sent to landfill

- at the heart of our strategy lies the need to tackle the growth in our waste

- we need to maximise the amount of value we recover from waste, through increased recycling, composting and energy recovery

- the strategy sets challenging targets for better waste management:

  – to recover value from 45% of municipal waste by 2010, at least 30% through recycling or composting

  – to recover value from two thirds of municipal waste by 2015, at least half of that through recycling and composting, and to go beyond this in the longer term

- we need to develop new and stronger markets for recycled materials – we will set up a major new programme, the sustainable waste action trust, to deliver more recycling and re-use, help deliver markets and end uses for secondary materials, and promote an integrated approach to resource use

- producers must increasingly expect to arrange for recovery of their products – in particular, we will develop an initiative on junk mail

- the amount of waste sent to landfill must be reduced substantially – we will introduce a system of tradable permits in England, restricting the amount of biodegradable municipal waste local authorities can send to landfill

- local authorities will need to make significant strides in recycling and composting – we will set statutory performance standards for local authority recycling and composting. We will work with local authorities to pilot schemes encouraging householders to reduce waste, and to participate in recycling schemes

- where energy recovery facilities are needed, we believe they should be appropriately sized to avoid competition with recycling, and the opportunities for incorporating Combined Heat and Power technology should always be considered

Waste and other Government policy initiatives

Waste has important links with many other aspects of Government policy. This strategy has therefore been prepared in light of the policies described in:

- The sustainable development strategy, *A Better Quality of Life*, May 1999 Cm 4345

- *Climate Change: Draft UK Programme*, DETR, February 2000, Product Code: 99EP0850. The Government's energy efficiency policy is being developed alongside the UK Climate Change Programme. The Government's energy policy, including sustainable energy, is set out in the DTI White Paper *Conclusions of the Review of Energy Sources for Power Generation and Government Response to Fourth and Fifth Reports of the Trade and Industry Committee* published in October 1998 (Cm 4071)

- *Sustainable Business – A consultation paper on sustainable development and business in the UK*, DETR, March 1998

- *Building a Better Quality of Life: A Strategy for more Sustainable Construction*, DETR, April 2000, Product Code: 99CD1065

- *New and Renewable Energy – Prospects for the 21st Century: Conclusions in Response to the Public Consultation Paper*, DTI, 2000

- *The Air Quality Strategy for England, Scotland, Wales and Northern Ireland*, DETR January 2000, Cm 4548, SE 2000-3, NIA-7

- *Sustainable Production and use of Chemicals – A Strategic Approach – The Government's Chemicals Strategy*, December 1999

- The Draft Soil Strategy – to be published for consultation later this year

- The White Paper on Integrated Transport, *A New Deal for Transport: Better for Everyone*, July 1998 Cm 3950

- the 2000 Budget statement

- and taking into account developments in local and regional government, including the Best Value initiative, and the establishment of Regional Development Agencies.

Application of this strategy

1.4 This waste strategy (both Part 1 and Part 2), together with guidance to planning authorities on the siting of facilities, implements for England and Wales the requirement within the Framework Directive on Waste[1], and associated Directives[2], to produce waste management plans. Strategies covering Scotland and Northern Ireland have also been prepared, by the Scottish Environmental Protection Agency and the Northern Ireland Environment and Heritage Service.

1.5 The requirements for waste management plans in these Directives is implemented into law by Section 44a of the Environmental Protection Act 1990 (as amended by the Environment Act 1995).

1.6 This strategy (both Part 1 and Part 2) is also a strategy for dealing with waste diverted from landfill in England and Wales, as required by the Landfill Directive[3].

1.7 This White Paper (both Part 1 and Part 2) replaces the previous waste management plan for England and Wales, published in June 1999 under the title *A Way With Waste – a draft waste strategy for England and Wales*.

1.8 Furthermore, this waste strategy is an advisory document. The 1990 Town and Country Planning Act requires local planning authorities in England and Wales to have regard to national policies in drawing up their development plans, and therefore this document will be an important source of guidance. These development plans will then provide a framework for individual planning decisions. Guidance on land use planning and waste in England is contained in Planning Policy Guidance notes 10 and 11. The equivalent documents in Wales are *Planning Guidance (Wales) Planning Policy First Revision April 1999*, and *draft Technical Advice Note on Planning, Pollution Control and Waste Management* issued in 1996.

---

1 This strategy describes the current policy on waste management in England and Wales. As such, it is a waste management plan under Council Directive 75/442/EEC as amended by Council Directive 91/156/EEC and adapted by Council Directive 96/350/EC (known as the Framework Directive on Waste)

2 Council Directive 91/689/EEC (the Hazardous Waste Directive), and the European Parliament and Council Directive 94/62/EC (the Packaging and Packaging Waste Directive)

3 1999/31/EC, known as the landfill Directive

CHAPTER 2

Identifying the problem

2.1 Identifying the size of the challenge facing us is essential before we can make decisions on how best to meet that challenge. This chapter includes details on how we define and categorise our waste. It also includes a detailed account of the quantities of waste currently produced in England and Wales.

Descriptions of waste

2.2 Determining what constitutes waste is not a simple affair. The legal definition of waste is discussed in Annex B section B.19 of this part of the strategy. Waste can also be divided into a number of different categories. A number of terms for waste are used throughout this strategy, and have the following meanings:

- **Controlled waste** – is used to describe waste which must be managed and disposed in line with waste management and other waste related regulations. It includes municipal, commercial and industrial waste. It can be from a house, school, hospital, shop, office, factory or any other trade or business. It may be solid or liquid; scrap metal, old newspapers, a used plastic bottle, etc. It does not need to be hazardous or toxic to be a controlled waste. Wastes collected from households, however, and certain animal wastes are specifically exempted from the Duty of Care requirements applied to other controlled wastes.

- **Municipal waste** – includes all waste under the control of local authorities or agents acting on their behalf. It includes all household waste, street litter, waste delivered to council recycling points, municipal parks and garden wastes, council office waste, civic amenity site waste, and some commercial waste from shops and smaller trading estates where local authority waste collection agreements are in place.

- **Household waste** – is defined in the Environmental Protection Act 1990, supplemented by the Controlled Waste Regulations 1992. It includes waste from household collection rounds, bulky waste collection, hazardous household waste collection and separate garden waste collection, plus waste from services such as street sweeping, litter and civic amenity sites. The definition also covers waste from schools.

- **Business (or commercial and industrial) waste** – covers commercial wastes and industrial wastes. Generally, businesses are expected to make their own arrangements for the collection, treatment and disposal of their wastes. Waste from smaller shops and trading estates where local authority waste collection agreements are in place will generally be treated as municipal waste.

- **Commercial waste** – waste arising from wholesalers, catering establishments, shops and offices.

- **Industrial waste** – waste arising from factories and industrial plants.

- *Agricultural waste* – is any waste from a farm or market garden, and includes organic matter such as manure, slurry, silage effluent and crop residues, but also includes packaging and films, and animal treatment dips (for example sheep dip).

- *Construction and demolition waste* – arises from the construction, repair, maintenance and demolition of buildings and structures. It mostly includes brick, concrete, hardcore, subsoil and topsoil, but it can also include quantities of timber, metal, plastics and (occasionally) special waste materials.

- *Mines and quarries waste* – includes materials such as *overburdon*, rock interbedded with the mineral, and residues left over from initial processing of the extracted material into saleable products.

2.3   In statistical terms, waste is defined by source, in terms of the Standard Industrial Classification (SIC) codes. Details of these codes can be found in the *Draft UK Waste Classification Scheme*.

Waste tracking and monitoring systems

2.4   There are a number of systems in place to monitor and track the movements of controlled waste. The Control of Pollution (Amendment) Act 1989 makes it a criminal offence for a person who is not a registered carrier to transport controlled waste to or from any place in Great Britain. The Controlled Waste (Registration of Carriers and Seizure of Vehicles) Regulations 1991 establish a system for registration of carriers of controlled waste and supplement the provisions of the Act dealing with the seizure and disposal of vehicles used for illegal waste disposal. The Regulations also set out the various groups who are exempt from the requirement to register.

2.5   Under the Environmental Protection (Duty of Care) Regulations 1991 (see Chapter 3 section 3.46 of this part of the strategy), when waste is passed from one person to another, the person taking the waste must have a written discription of it (unless one of the parties is exempt from the Duty of Care). A transfer note must also be filled in and signed by both persons involved in the transfer. The transfer note will include a range of information on the waste, such as what the waste is and how much there is of it what sort of container it is in, where the transfer took place, and details of the people involved in the transfer.

2.6   The movements of special waste are tightly controlled by a system of consignment notes introduced as part of the Special Waste Regulations 1996 (see Chapter 6 of this part of the strategy).

2.7   The policy and practicalities of importing or exporting waste are covered in the *United Kingdom Management Plan for Exports and Imports of Waste*[1], which is currently being reviewed. In general, the importing or exporting of waste for disposal is banned (with a few specified exemptions, such as importing certain wastes for high temperature incineration), while the import and export of waste for recovery is permitted in certain circumstances.

2.8   There are also obligations on certain businesses to keep track of the amount of packaging they produce, so that they comply with the Producer Responsibility Obligations (Packaging Waste) Regulations 1997.

1 United Kingdom Management Plan for Exports and Imports of Waste, HMSO, £15.00, ISBN 0-11-753181-2

Exports and imports of waste

The *1989 United Nations Basel Convention on the control of transboundary movements of hazardous wastes and their disposal* provides the framework for a global system of controls on international movements of hazardous and certain other wastes. Countries who sign up to the Convention are obliged to take appropriate measures to:

- reduce transboundary movements to a minimum, consistent with their environmentally sound management

- minimise the generation of waste

- aim at self-sufficiency in final disposal

- prevent transboundary movements where the proposed country of destination has not given its consent

- ensure environmentally sound disposal and recovery of wastes

- (subject to the ratification of the "Ban" amendment agreed in 1995) prohibit the export of hazardous waste from Annex VII countries (for example those belonging to the OECD, the European Union and Liechtenstein) to non-Annex VII countries.

The United Kingdom, together with the European Commission and most other EC Member States, became a Party to the Convention in May 1994, and implemented the provisions of the Convention in the European Community by way of Council Regulation (EEC) No. 259/93 (the *Waste Shipments Regulation*). This was amended by Council Regulation 120/97 to implement the ban on exports of hazardous wastes from EU Member States to non-OECD Countries.

The Waste Shipments Regulation also incorporates the OECD *Decision on the Control of Transfrontier Movements of Waste Destined for Recovery Operations*. This Decision, which is regarded as a multilateral agreement under the Basel Convention, was introduced to facilitate movements of wastes for recovery between countries belonging to the OECD.

Within this framework, the United Kingdom has some scope to draw up additional controls on shipments of waste. Such controls are set out in the legally binding *United Kingdom Management Plan for Exports and Imports of Waste* (the UK Plan). The UK Plan reflects long-standing UK policies of self-sufficiency in waste disposal and the proximity principle, whereby waste should be disposed of in, or as close as possible to, the country of origin. It therefore prohibits exports and imports of waste for disposal, except in limited circumstances. At the same time, the UK Plan seeks to preserve the trade in wastes for genuine and environmentally sound recovery operations in line with international agreements.

The UK Government recently launched a public consultation exercise on future policies on waste exports and imports. It is intended that the policies proposed in the revised UK Plan should complement those in this strategy. The draft revised UK Plan recommends maintaining the existing policy framework, but with minor changes to reflect operational experience and other political and environmental developments.

Commercial and industrial waste data

2.9 The Environment Agency has undertaken a survey of industrial and commercial waste, providing detailed information of the amounts and different types of waste produced by industrial and commercial companies in England and Wales.

2.10 Data was collected by means of site visits and by telephone, from a sample of about 20,000 companies between October 1998 and April 1999, and a full analysis of the data from the survey will be published by the Agency in 2000.

2.11 A summary of the provisional data is given below. Further work is continuing in order to improve the conversion factors used to convert waste volumes to waste tonnages, and there will be some changes to the figures when the final conversion factors have been calculated. The preliminary results show:

- total industrial waste is estimated at 48 million tonnes
- total commercial waste is estimated at 30 million tonnes

Estimated totals for different sectors of industry and commerce

Business sector	Estimated annual waste generation Million Tonnes
Industrial companies	
Food, drink and tobacco	8
Textiles, wood, paper	7
Chemicals, rubber, mineral products	9
Metals, metal products	8
Other manufacturing	7
Coke, oil, gas, electricity, water	3
Transport, storage, communications	4
Miscellaneous	2
Total industrial	**48**
Commercial companies	
Wholesale	4
Retail	7
Hotels and catering	4
Education	2
Other business and public administration	13
Total commercial	**30**
Total industrial and commercial	**78**

Estimated totals for identified waste streams

Waste type	Estimated annual waste generation Million Tonnes
Inert, in-house (small scale) construction	2
Paper and card	7
Food	3
Other general and biodegradable	9
Metals and scrap equipment	6
Contaminated and healthcare	5
Mineral waste and residues	6
Chemicals	4
General commercial	23
General industrial	13
Total	**78**

Recovery and disposal routes for each identified waste type, by percentage

Waste type	Waste recovered: recycled	other	total	Waste disposed: landfill	other	total
Inert, in-house construction	39%	0%	39%	56%	5%	61%
Paper and card	76%	1%	77%	22%	1%	23%
Food	69%	11%	80%	7%	13%	20%
Other general and biodegradable	42%	21%	63%	26%	11%	37%
Metals and scrap equipment	89%	0%	89%	10%	1%	11%
Contaminated and healthcare	34%	2%	36%	42%	22%	64%
Mineral waste and residues	38%	0%	38%	62%	0%	62%
Chemicals	21%	7%	28%	45%	27%	72%
General commercial	18%	4%	22%	78%	0%	78%
General industrial	11%	2%	13%	86%	1%	87%

Recovery and disposal routes (million tonnes – calculated from data given above)

Waste type	Waste recovered: recycled	other	total	Waste disposed: landfill	other	total
Inert, in-house construction	0.78	0.00	0.78	1.12	0.10	1.22
Paper and card	5.32	0.07	5.39	1.54	0.07	1.61
Food	2.07	0.33	2.40	0.21	0.39	0.60
Other general and biodegradable	3.78	1.89	5.67	2.34	0.99	3.33
Metals and scrap equipment	5.34	0.00	5.34	0.60	0.06	0.66
Contaminated and healthcare	1.70	0.10	1.80	2.10	1.10	3.20
Mineral waste and residues	2.28	0.00	2.28	3.72	0.00	3.72
Chemicals	0.84	0.28	1.12	1.80	1.08	2.88
General commercial	4.14	0.92	5.06	17.94	0.00	17.94
General industrial	1.43	0.26	1.69	11.18	0.13	11.31
Total commercial and industrial	**27.68**	**3.85**	**31.53**	**42.55**	**3.92**	**46.47**
Total percentages	35.49%	4.94%	40.42%	54.55%	5.03%	59.58%

2.12   Overall, about 40% of industrial and commercial waste is recovered (35% being recycled), and the other 60% is disposed, mainly to landfill. Nearly half the total waste is identified as general commercial or industrial waste, of which about 80% is landfilled and about 15% is recycled. Recycling is highest for separately collected waste streams such as metals and scrap equipment (89%) and paper and card (76%).

2.13   Analysis of the data from the survey will provide – for the first time – detailed information on the amounts of different types of waste produced by companies in each industrial and commercial sector, together with details of treatment and disposal routes for the waste. The 1998/99 survey will establish baseline data for industrial and commercial waste. The Agency plans to carry out further surveys every three years to update the information. It is planned to make the results of these surveys widely accessible through a waste database:

- a report of the survey will be published in 2000

- strategic waste management assessment reports will also be published in 2000 for each planning region in England and Wales

- results of the survey will also be made available through a website which will include a suite of interactive tools, including a benchmarking tool for businesses, a waste exchange and a recycling database

Business environmental performance

Alongside factual questions such as types and quantities of wastes generated, and methods of waste management used, companies were also asked to assess their own environmental performance. While the responses to this question are not statistically representative of business environmental performance, the results do offer some indication of how businesses generally view their own performance on waste management issues.

Some of the salient points to emerge include:

- 16% of businesses surveyed claim to have achieved waste reduction (mainly as a result of carrying out waste audits) and a further 2% plan to achieve waste reductions during the next 12 months

- 40% of companies claim to monitor waste expenditure and a further 3.5% plan to introduce waste expenditure monitoring in the next 12 months

- 20% of companies report having carried out a waste audit and 60% of those report waste savings as a result of the audit

- 30% of companies claim to have a published environmental policy, 80% of which either include waste policies or cover waste issues; a further 5% of companies plan to publish an environmental policy in the next 12 months

- 10% of companies are members of a local Business Environment Group and a further 2% plan to join one in the next 12 months.

Municipal waste data

2.14 Starting in 1995/96, the Department of the Environment, Transport and the Regions (DETR) and the Welsh Office (the National Assembly for Wales from 1 July 1999) have commissioned an annual survey of local authorities in England and Wales, to collect information on the collection, treatment and disposal of municipal and household waste. Local authorities' response to the survey has been very good, and information was provided by 90% of authorities in 1995/96, rising to 95% in 1997/98. So far, all but one local authority have replied to the 1998/99 survey.

2.15 This high response has enabled reliable national estimates to be made of levels of municipal and household waste; recycling rates; and methods of waste collection, disposal and treatment routes. The survey also collects information on the amounts of waste collected for centralised composting and asks local authorities to provide estimates of the amounts diverted to home composting. Information from the first three years of the survey show that:

- There were around 27 million tonnes of municipal waste in 1997/98, up from 25.2[2] million tonnes in 1995/96

2 The figure for total municipal waste has been revised from 25.9 million tonnes since the publication of Municipal Waste Management 1995/96. The revised figure takes account of late returns, and uses information from subsequent surveys to make improvements to the grossing methodology.

- over 90% of municipal waste comes from household sources – 24.6 million tonnes in 1997/98 – which represents about 22 kg per household per week

- the majority of municipal waste, 85%, was disposed of to landfill in 1997/98, slightly more than in 1995/96

- in 1997/98, 14% of municipal waste had value recovered from it through recycling, composting or energy from waste schemes, up from 12% in 1995/96

- around two million tonnes of household waste were collected for recycling or composting in 1997/98

- the percentage of household waste collected for recycling or composting increased from 6.5 per cent in 1995/96 to 8% in 1997/98

- Paper and card accounted for nearly 40% of the household waste collected for recycling in 1997/98, while glass and centralised composting accounted for another 20% each

- the number of *bring* sites and civic amenity sites for recycling has increased slightly to a level of about eight sites per ten thousand households, but there has been a substantial increase in the number of households served by *kerbside* recycling schemes, from 17% of households in 1995/96 to 37% in 1997/98

- amounts of waste collected for centralised composting have more than doubled in two years, to 390,000 tonnes in 1997/98; in addition, local authority estimates suggest that between 200,000 and 300,000 tonnes of waste are composted at home

- the use of wheeled bins for the collection of waste from households has increased from 38% of households in 1995/96 to 42% in 1997/98

2.16 The preliminary results of the 1998/99 survey show that:

- there were around 27.9 million tonnes of municipal waste arising in 1998/99

- of this total around 83% was landfilled, while 9% (over 2.5 million tonnes) was recycled and 8% was incinerated with energy recovery, giving a total municipal recovery figure of 17%.

Sustainable waste and resource use indicators

2.17 Following the publication of the Government's Sustainable Development Strategy, *A better quality of life*[3], launched by the Deputy Prime Minister in May 1999, a baseline assessment of a series of headline and core indicators for sustainable development was set out in *Quality of life counts*[4], published in December 1999.

3 *A better quality of life* – a strategy for sustainable development for the UK, published by The Stationery Office, May 1999. ISBN 0-10-143452-9. Price £11.80.
4 *Quality of life counts* – indicators for a strategy for sustainable development for the United Kingdom: a baseline assessment, published by the Government Statistical Service, 1999. ISBN 1-851123-43-1. Price £22.00.

2.18 A number of these indicators (which cover the whole of the UK, not just England and Wales) measure resource use and waste management. These baseline indicators are reproduced in full below. Note that these indicators do not take into account the Environment Agency's recent survey of industrial and commercial waste. The text below is reproduced from *Quality of Life Counts* and does not reflect the policy developments in this strategy.

WASTE ARISINGS AND MANAGEMENT (Indicator H15)

Objective   **Move away from disposal of waste towards waste reduction, reuse, recycling and recovery**

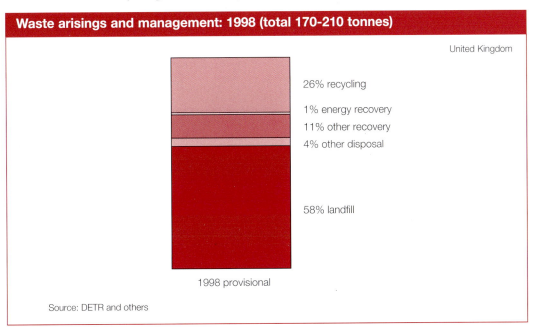

It is estimated that between 170 and 210 million tonnes of waste are produced each year in the UK by households, commerce and industry, including construction and demolition. Nearly 60 per cent of this is disposed of in landfill sites.

Relevance   The types of waste we produce, all forms of waste management, and the transport of waste, have impacts on the environment. Waste is a potential resource and increased levels of reuse, recycling and energy recovery will contribute to achieving more sustainable lifestyles.

Targets and goals   Range of targets in draft waste strategies for England and Wales, Scotland and Northern Ireland.

Trends   The 1998 estimate is provisional, and will be revised when final information from current surveys is available. For most sectors there are no comparable data for earlier years. Trends in household waste are illustrated in Indicator A5, which shows an increase of 26% in total household waste in England and Wales between 1983/84 and 1997/98.

Background   The Government and the National Assembly for Wales are committed to achieving targets derived from European legislation, such as the Landfill

Directive and the Packaging Directive. The Landfill Directive, which requires substantial amounts of waste to be diverted from landfill, will require a step change in the management of municipal waste in the UK.

Strategies *A better quality of life*: a strategy for sustainable development in the UK; *A Sustainable Wales: Learning to Live Differently* – a draft Sustainable Development Scheme for Wales; *National waste strategy: Scotland*; *Waste management strategy 1999-2019 – Northern Ireland*.

UK RESOURCE USE (Indicator A1)

Objective **Greater resource efficiency**

An indicator showing, for example, UK consumption of materials by weight or volume per unit of Gross Domestic Product, and identifying broad resource groups separately, such as metals, fossil fuels, minerals and renewables (for example cereals, timber).

Relevance A key sustainable development objective is to use natural resources more efficiently. The rate of consumption of resources should not reduce their availability for future generations, and producing more with less means reducing environmental pollution and degradation caused by the extraction, use and disposal of natural resources.

Targets and goals There are no specific resource efficiency targets for the UK but there is a commitment to promote continual improvements in resource efficiency.

Background Specific aspects of efficiency are dealt with by other indicators such as energy efficiency of the economy, and competitiveness/ productivity.

Research will be carried out to identify suitable measures, consistent as far as possible with work which is being carried out by other countries and international bodies such as the Organisation for Economic Cooperation and Development (OECD) and the United Nations (UN) to produce first estimates for the UK.

One approach is to estimate Total Material Requirements (TMR). This looks at the use of resources, in weight or volume terms, for broad resource groups; identifies separately, direct material inputs (both renewable and non-renewable natural resources), and hidden or ancillary flows such as excavated or disturbed material from mining; and identifies consumption using imported materials.

WASTE BY SECTOR (Indicator A4)

Objective **Move away from disposal of waste towards waste reduction, reuse, recycling and recovery**

An indicator showing waste produced by different sectors, possibly in relation to GDP or output. The indicator will be developed when final figures are available from the Environment Agency's survey of industrial and commercial waste.

Relevance   Sustainable development requires an improvement in resource efficiency. Waste reduction will contribute to this.

Targets and goals   The draft waste strategy for England and Wales *A Way With Waste*, and the draft waste strategy for Northern Ireland *Waste Management Strategy 1999 – 2019*, both propose a target of reducing industrial and commercial waste to landfill to 85% of 1998 levels by 2005. The draft waste strategy for Scotland *National Waste Strategy: Scotland* proposes a target of reducing industrial waste arisings by 3-5% by 2005.

Background   In the UK around 400 million tonnes of waste a year are produced from different sources. About 20-25% of this is from industry and commerce.

HOUSEHOLD WASTE AND RECYCLING (Indicator A5)

Objective   **Move away from disposal of waste towards waste reduction, reuse, recycling and recovery**

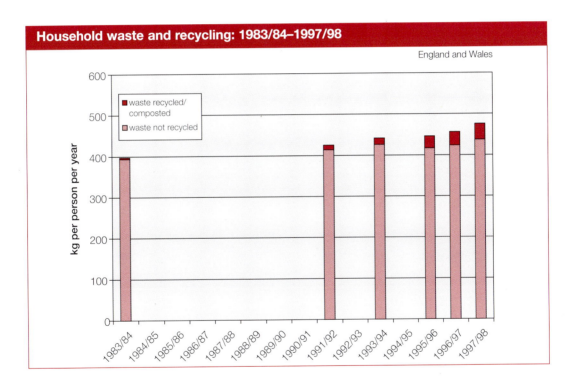

In England and Wales, amounts of household waste generated have increased steadily to nearly 500kg per person per year in 1997/98. This represents an increase of 26% in total household waste and 20% in household waste per head. About 8% of this waste is recycled or composted.

Relevance   Household waste reduction and increased recycling would lead to a reduction in the environmental impact of waste disposal.

Targets and goals   The draft waste strategy for England and Wales *A Way With Waste* sets a goal of 30% recycling and composting by 2010. Options for introducing a household waste reduction target will be considered. The draft waste management strategy for Northern Ireland sets a target of 25% recycling or composting by 2010, and also includes a target of reducing household waste to 1998 levels by 2005 and thereafter by at least 1% annually. The Scottish draft waste strategy proposes a target for a 2-4% reduction in municipal waste between 1994 and 2016.

Trends

It is difficult to compare long-term changes because of differences in data sources and definitions. The increase in levels of household waste is likely to be linked to a number of factors, including the increase in number of households and changes in the pattern of consumer spending. There may also be an increase in the amount of commercial waste mixed in with household waste. Improved recycling rates reflect improved provision of recycling facilities.

Background

Household waste includes household bin waste and also waste from civic amenity sites, other household collections, recycling sites, litter collections and street sweeping. Household waste represents about 90% of municipal waste, which is collected and managed by local authorities. Most recycling of household waste comes from "bring" sites such as bottle and paper banks, and increasingly from kerbside collections.

MATERIALS RECYCLING (Indicator A6)

Objective

Move away from disposal of waste towards waste reduction, reuse, recycling and recovery

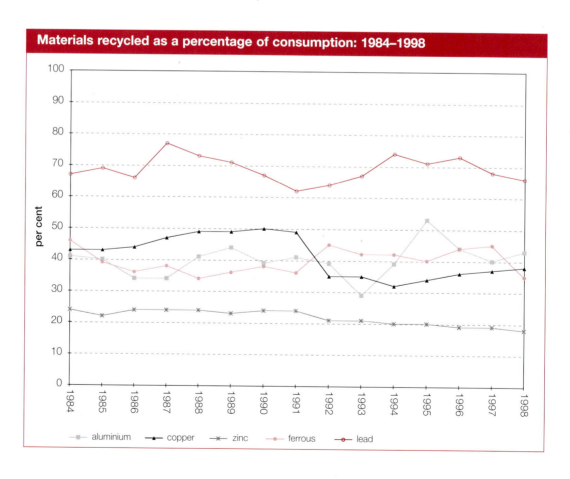

Materials recycled as a percentage of consumption: 1984–1998

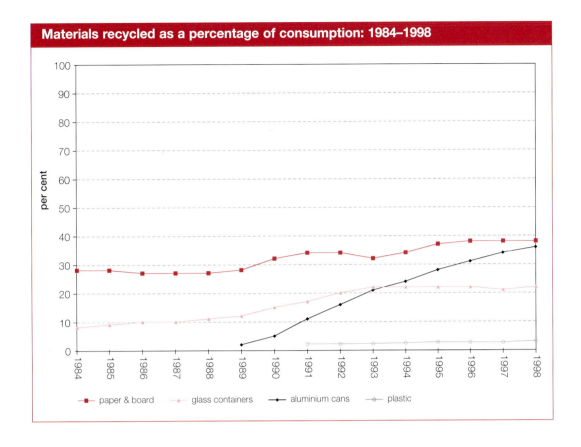

Materials recycled as a percentage of consumption: 1984–1998

Legend: paper & board, glass containers, aluminium cans, plastic

The average recycling rate for metals has been fairly stable between 1984 and 1998 at around 40%. In 1998, about 40% of paper was also recycled, but the level of glass recycling is lower, at around 22%, and only 3% of plastics are recycled.

Relevance	Increased levels of recycling ensure that waste is used as a resource, and value is obtained from it.
Trends	Recycling rates for industrial process waste are generally high. Recycling of paper increased steadily between the mid-1980s and mid-1990s, encouraged by improved recycling facilities, but showed no further increase up to 1998. After earlier improvements, the rate of glass recycling has also stabilised.
Background	For many materials, the scope for recycling is limited by the size of the markets for secondary materials. The level of recycling can also be constrained by lack of clear standards, poor consumer awareness and the volatility of prices for recycled materials.

HAZARDOUS WASTE (Indicator A7)

Objective	**Move away from disposal of waste towards waste reduction, reuse, recycling and recovery**

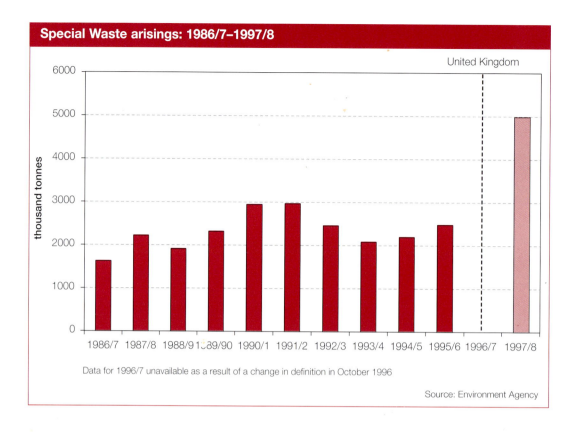

Special Waste arisings: 1986/7–1997/8

United Kingdom

thousand tonnes

Data for 1996/7 unavailable as a result of a change in definition in October 1996

Source: Environment Agency

Up to 1995/96, around 2 to 3 million tonnes of special waste were reported each year. Following an extension of the definition in 1996 to include some further waste types, such as waste oils, total special waste was around 5 million tonnes in 1997/98.

Relevance	Managing and disposing of hazardous waste has a particularly high impact on the environment.
Trends	Amounts fluctuate from year to year, partly due to variations in amounts of contaminated soil removed for remediation purposes. There is no clear trend.
Background	Before 1996, special waste was defined by the Control of Pollution (Special Waste) Regulations 1980. The Special Waste Regulations 1996 defined a wider range of hazardous wastes as special. Following the introduction of these regulations, all movements of special waste are tracked until they reach a waste management facility.
	It is anticipated that further changes will be made to the list of special wastes over the next few years. These additions may substantially increase the tonnage of wastes defined as special, regardless of trends in the overall volume of waste generated. The amount of special waste produced will also be affected by any new measures taken to remove hazardous chemicals from the utility chain. Further work will be needed to develop this indicator so that it reflects trends in the amount of hazardous waste generated, independent of the definition of special waste.

Waste research, advice and information sources

2.19 The Government and the National Assembly recognise the need to work in partnership with a wide range of othe groups, if we are to achieve the vision, objectives and targets set out in the Waste Strategy. Part of this dialogue with other groups includes coordinating and rationalising our mutual efforts on waste research, and promotion and information campaigns.

2.20 The move towards sustainable waste management relies on our ability to make rational decisions about the Best Practicable Environmental Option for local waste at the local level. The ability to make the best decision depends on having good data on which to base the decision-making process. Decisions will also be influenced by facts determined and underpinned by scientific research. Furthermore, it is not enough to obtain the data and undertake the research. The results of these efforts need to be disseminated to all those organisations and communities involved in making decisions on waste. Information and promotion campaigns become essential once a local sustainable waste management strategy has been agreed on and implemented – everybody needs to know what actions they need to take in support of the new system.

2.21 The Waste Action and Resources Programme, the new sustainable waste management programme, is designed to coordinate data generation and dissemination, and will have the resources to initiate research, and to offer advice on re-use and recycling best practice. But the new programme will not replace existing activities on waste research and promoting waste-awareness – rather it will complement and extend these existing initiatives.

WASTE AND RESOURCES ACTION PROGRAMME

2.22 The Government and the National Assembly are determined to overcome the market barriers to promoting the re-use and recycling of waste materials. To that end, a new programme – the Waste and Resources Action Programme – is being established and developed in 2000, to promote more sustainable waste management through an integrated approach to materials resource use, extensive development of markets and end-uses for secondary materials.

2.23 The central goals of the Programme will be to:

- achieve a significant increase in waste reduction and re-use, to help meet the vision of the waste strategy

- double the present recycling and composting rates, to help meet the targets and goals set out in the strategy

2.24 Both in its ethos and its structure, the Programme will take a partnership approach involving DETR, DTI, the devolved administrations and the private sector, with contributions from each partner. In order to achieve its goals, the Programme will need to quickly develop its expertise and information management capabilities, and draw up plans for commissioning research and development to support its activities.

2.25 Further details of the Programme can be found in Chapter 3 of Part 1 of this strategy.

OTHER WASTE RESEARCH INITIATIVES

2.26 Research plays an important part in our understanding of waste issues and a number of key organisations undertake and support research into waste related issues. Within Government waste related research is carried out by DETR, DTI, MAFF and the Environment Agency. The National Assembly for Wales also undertakes research.

2.27 The Department is aware that there are a number of organisations who also promote research through Environmental Bodies, these include the Environmental Services Association's research trust ESART and the new body set up by the Institute of Wastes Management. To ensure that DETR's research does not duplicate such activity it will be considering the aims and objectives of its own programme and links with others.

2.28 The current objectives of DETR's waste research programme are to support:

- the development of sustainable waste management with an emphasis on reducing waste, optimising re-use and recovery of materials

- the development of an integrated strategy for waste management

- measures to ensure that waste is properly controlled so as to protect human health and the environment

- our waste interests internationally

DISSEMINATION OF WASTE FACTS, INFORMATION AND BEST PRACTICE

2.29 Informing businesses, consumers and waste stakeholders is a vital element of the waste strategy. Ensuring that the messages being disseminated do not conflict with each other means that Government dissemination and promotion strategies need to be discussed and (when possible) coordinated with the efforts of other groups.

2.30 In particular, DETR is involved in three waste related promotions in 2000:

- *'are you doing your bit?'* – DETR through its *'are you doing your bit?'* campaign is encouraging people to take on board environmental considerations in their every day lives. The aim is to encourage people to understand that small actions undertaken by them can make a difference, thus overcoming the feeling that a number of environmental issues are too large for them to do anything about. Waste is one of the themes of the 2000/01 campaign. Adverts will appear in magazines and newspapers, and on television. Messages include: re-use; recycling and buy recycled.

- *National Waste Awareness Initiative* – seeks to raise awareness about waste issues and to create long term changes in attitudes and behaviour in the way waste is dealt with across the UK. The aim is to persuade people to take more responsibility and ownership of the waste they create and to deal with it in ways that are more sustainable. The Initiative is planning to launch its image and key campaign messages during 2000, and aims to start the campaign itself in 2001. DETR sees this as an important initiative and in recognition of that has provided funding for administration of the Initiative. The Government would like to see clear links between the Initiative and *are you doing your bit?*

- *Going for Green Waste Theme Month* – The aim of this particular initiative is to conduct a month long campaign (in October) to raise the profile on issues around the need to reduce, re-use and recycle waste. Activities are undertaken in conjunction with partners especially local authorities. Activities undertaken include national and regional events, a children's Activity Book, supporting newspaper articles and a national seminar at the end of the month.

CHAPTER 3

The decision-making framework

3.1 The Government and the National Assembly for Wales believe that the most effective waste management decisions can be taken by adopting an integrated approach to waste management. Integrated waste management can be considered to be a number of key elements working in concert, in particular:

- *recognising each step in the waste management process as part of a whole* – decisions should take account of the collection, transport, sorting, processing and recovery or disposal of wastes; and in the case of recovery, identification of end uses or markets for the resulting materials and energy

- *involvement of all key players* – an integrated approach to waste management should also define the contributions which all interested parties (which might include waste producers and managers, waste reprocessors, waste regulators, waste management planners, community groups, consumers and householders, and Government) can make in the development and achievement of common goals and objectives

- *a mixture of waste management options* – those planning the management of significant quantities of various wastes should avoid over-reliance on a single waste management option. It is unlikely that one approach will represent the Best Practicable Environmental Option (BPEO) for all elements of the waste stream

- *formal and informal partnerships* – especially between those organisations obligated with legal responsibilities for managing waste that they generate or that arises in their areas. In particular this means Waste Collection and Waste Disposal Authorities within a particular area. Local authorities within a region who will also need to take a collective view of the more strategic regional implications of their various policies towards waste management issues

Hampshire's Project Integra

Councils in Hampshire have recently been awarded Beacon Council status for their work on integrated waste management. The councils, in partnership with the private sector, are reducing dependency on landfill through a combination of recycling and composting techniques, planned energy from waste, and a strong emphasis on community consultation and education on waste issues.

The consortium has invested significant sums in research into waste analysis and public attitude surveys. Clear development plans exist for the expansion of recycling infrastructure with an aim of recycling 40% of municipal waste, alongside the provision of energy from waste capacity. The project partners have also displayed a welcome emphasis on disseminating good practice in the process of developing integrated waste strategies, and a willingness to share failures as well as successes.

3.2 The technique that should be used for making waste management decisions is known as Best Practicable Environmental Option (BPEO), and the simplest way to encourage integrated waste management is to structure the implementation of BPEO with the above key elements in mind. The following section gives some further detail on determining BPEO. Subsequent sections in this Chapter consider the impact of land use planning, waste management licensing, integrated pollution prevention and control regulation, and the Duty of Care, which all have a role to play in determining and implementing optimum waste management solutions across England and Wales. Decisions on how to treat or dispose of waste should be taken locally, taking account of local circumstances and the nature of particular waste streams. When taking waste management decisions on suitable treatment options, sites and installations, local authorities must follow the framework set out below. This framework should also act as a guide for other decision-makers, including business waste managers.

The precautionary principle

Any integrated waste management system must make allowance for the precautionary principle, which states that where there are threats of serious or irreversible damage, lack of full scientific certainty shall not be used as a reason for postponing cost-effective measures to prevent environmental degredation.

Determining the Best Practicable Environmental Option

3.3 If we are to manage our waste more sustainably, decision makers need the tools to move us in that direction. Waste is not a single substance, and its management is not a series of simple choices. Rather it is, for the most part, a complex mixture of different materials, in differing proportions. Each of these materials has the potential to impact differently on the environment. Environmental impact can also be influenced by the collection system used, the locations where waste is generated, managed and disposed of, and the resources consumed through managing our waste. In a sustainable and integrated system all these factors must be taken into account when making decisions on how best to manage waste.

3.4 The process that should be used for considering the relative merits of various waste management options in a particular situation is the Best Practicable Environmental Option (BPEO). This was defined in the 12th Report of the Royal Commission on Environmental Pollution as:

> *the outcome of a systematic and consultative decision-making procedure which emphasises the protection and conservation of the environment across land, air and water. The BPEO procedure establishes, for a given set of objectives, the option that provides the most benefits or the least damage to the environment as a whole, at acceptable cost, in the long term as well as in the short term.*

3.5 We have long sought to protect the local environment and human health from the adverse effects of waste management through a comprehensive system of planning and pollution control legislation. Sustainable development challenges us to develop more integrated systems for managing waste that are environmentally effective (both locally and globally), economically affordable and socially acceptable.

3.6   When considering the BPEO, decision makers need to have regard to international obligations (such as the biodegradable municipal waste diversion targets in the Landfill Directive), the national policy framework as set out in this strategy (including the waste hierarchy), and policy guidance at regional and local level. The concept of BPEO means that local environmental, social and economic preferences will be important in any decision. These may well result in different BPEOs for the same waste in different areas, or even different BPEOs for the same type of waste in the same area but at different times (for example, when the economy is growing or in recession).

The proximity principle

The proximity principle suggests that waste should generally be disposed of as near to its place of origin as possible. This is in part to ensure that we do not simply export problems to other regions or countries. It also involves recognition that the transportation of wastes can have a significant environmental impact. A network of facilities would enable these environmental impacts – and in many cases financial costs – to be reduced.

The proximity principle has two important functions:

- it is a tool for planning authorities and businesses when considering the requirements for, and location of, waste management facilities and regional self-sufficiency

- it helps raise awareness in local communities that the waste they produce is a problem with which they must deal

The proximity principle can make the link between the waste hierarchy and BPEO. Where the BPEO for a waste stream is towards the lower end of the waste hierarchy, this can often be because the environmental impact or cost of transport to a distant reprocessing facility or market outweighs the benefit of recovering the waste. Planners should consider the mode of transport and not just the distance: a longer journey by river or rail may be environmentally preferable to a shorter road journey.

In some respects, the proximity principle is particularly applicable to hazardous wastes, as they are intrinsically hazardous and moving them over long distances may increase the risk of damage arising.

However, it is also important that hazardous wastes are dealt with at a facility at which they can be treated in an environmentally sound manner. Because of the relatively low level of arisings of some of these wastes, there are likely to be relatively few suitable facilities for their disposal. Thus, the need for appropriate treatment should be considered alongside the proximity principle when considering where hazardous wastes should be disposed of.

It is therefore important that waste planning authorities and businesses consider the need for a network of specialised disposal facilities for hazardous wastes produced across England and Wales, and collaborate accordingly. This will become even more important with the implementation of the provisions of the Landfill Directive which will limit the disposal of hazardous wastes to landfill.

3.7   The waste hierarchy provides a theoretical framework which should be used as a guide for ranking the waste management options being considered as part of the BPEO assessment. It offers an order which can be used when considering various waste management options, starting with a review of how less waste might be produced. Once this has been carried out, all options in the hierarchy should be considered for each component material within the waste stream, and for waste which cannot be reasonably separated out. For different materials, different options are likely to prove more environmentally effective and economically affordable. Thus the BPEO for a waste stream is likely to be a mix of different waste management methods.

The waste hierarchy

The waste hierarchy is a conceptual framework, which acts as a guide to the framework that should be considered when assessing BPEO. It can also be a useful presentational tool for delivering a complex message in a comparatively simple and accessible way:

- the most effective environmental solution is often to reduce the generation of waste – *reduction*
- products and materials can sometimes be used again, for the same or a different purpose – *re-use*
- value can often be recovered from waste, through *recycling*, *composting* or *energy recovered*
- only if none of the above offer an appropriate solution should waste be disposed of.

3.8   To make rational decisions on waste management, we need to consider a number of subsidiary objectives. These might include social, economic, environmental, land use, and resource use impacts, each of which will have its own set of criteria.

3.9   The judgement about which mix of waste management options provides the BPEO can be resolved by analysing the trade-offs between objectives or criteria. This can show the extent to which one objective is sacrificed in order to achieve another (for example, how much costs could rise to reduce the impact on global warming). Formally, this can be resolved using decision techniques such as *multi-criteria analysis* (MCA). These entail the systematic modelling of decision-makers' preferences, to resolve the choice between several options involving a number of objectives or criteria. By aggregating disparate information onto a common index of value they provide a rational basis for classifying choices.

3.10   Even where such a formal methodology is not adopted, there are considerable advantages in using an approach that is:

- *comprehensive*: ensuring that all concerns regarding waste management alternatives can be seen to have been identified and addressed

- *flexible*: allowing the robustness of potential decisions to be thoroughly explored

- *iterative*: enabling development and refinement of the options

- *transparent*: so that the reasons behind a particular choice are made clear

Identifying waste management options – step-by-step

Identifying the most sustainable mix of waste management options, environmentally, economically and socially, can be a daunting task. However, the process can be simplified by breaking it down into smaller, more manageable tasks:

Step 1:   set the overall goals for making the waste management decision, subsidiary objectives and the criteria against which the performance of different options will be measured

Step 2:   identify all the viable options

Step 3:   assess the performance of these options against the criteria

Step 4:   value performance

Step 5:   balance the different objectives or criteria against one another

Step 6:   evaluate and rank the different options

Step 7:   analyse how sensitive the results are to variations in the assumptions made or the data used.

LIFE CYCLE ASSESSMENT

3.11 Determining the financial costs of waste management alternatives is relatively straightforward, but assessing environmental and social performance is much more complex. Indicators can include the conservation of non-renewable resources, release of greenhouse gases, emissions which may impact on air quality, noise, visual intrusion and traffic congestion. All could have an impact on communities, the local economy or the environment.

3.12 A further complication is that our choice of waste management options can have a substantial impact outside the waste management system. This could be, for example:

- changes in the amount of a particular fuel consumed in power stations and a consequent change in pollution levels, as a result of decisions made about implementing waste management options which include generating electricity from waste

- a decision to recycle aluminium cans will have impacts on the bauxite mining industry, the aluminium processing industry, and on the energy and resource use by each affected sector

3.13 To find an overall, optimal, environmental solution for managing waste, without the risk that our decision will result in a worsening of the overall impact, we need to adopt a life cycle approach. Life cycle assessment (LCA) is the systematic identification and evaluation of all the environmental benefits and disbenefits that result, both directly and indirectly, from a product or function throughout its entire life – from extraction of raw materials to its eventual disposal and assimilation into the environment.

3.14 Life cycle assessment can provide a basis for making strategic decisions on the ways in which particular wastes in a given set of circumstances can be most effectively managed. Even where a comparison of different systems does not show a clearly preferred option in terms of quantifiable environmental flows, this indication of environmental performance can be of value to decision-makers.

3.15 Life cycle assessment also takes account of the proximity principle – that waste should be dealt with as close to the point of its generation as practically possible. The transportation of waste (both in terms of distance travelled and the mode of transportation) from the point at which waste is generated, through the collection and sorting of waste, to where it is treated, recovered or finally disposed are included within the life cycle.

Life cycle assessment tools

3.16 Life cycle assessment can be a time-consuming process. The Environment Agency (together with the Scottish Environmental Protection Agency and the Northern Ireland Environment and Heritage Service) is carrying out a programme of research into the environmental burdens, and related impacts, of waste management options from cradle to grave. Data has been collected on the environmental flows (known as *burdens*) of household waste associated with each of the following key areas of waste management:

- Waste transport and other vehicle use
- Waste collection and separation
- Incineration

- Landfill
- Composting and anaerobic digestion
- Recycling of materials

3.17 The combined inventory of environmental data encompasses the burdens associated with all operations undertaken between the point of waste production and its ultimate reduction to inert material, including any burdens that can be offset against materials and energy recovery.

3.18 The inventories of data are being made publicly available, and the Agencies have produced a software tool (WISARD) which incorporates the data, to enable waste managers (in local authorities and in businesses) to model waste management systems from components representing individual operations. The software will also calculate the environmental flows and the associated impacts resulting from different parts of the system, and allow different systems to be compared for their relative environmental performance.

3.19 As well as the Agencies' software, others have produced tools applying life cycle techniques to waste management in the UK. The Integrated Waste Management tool[1] allows the user to develop life cycle inventories of municipal waste management systems in the same way as WISARD. A much improved version of this software is expected to be publicly available in 2000.

EXTENDING THE RATIONALE OF DECISIONS

3.20 Although the life cycle assessment approach ensures all quantifiable impacts are taken into account, the concept does not address all of the criteria that need to be taken into account in any decision-making process. To extend beyond the present boundaries, the Government has begun research into developing a structured framework for waste planning authorities that takes into account wider environmental and social factors. The outcome of this research is expected to be available by early 2002 and will, together with life cycle assessment, help decision-makers identify the most appropriate waste management options.

Land use planning and waste

3.21 The Town and Country Planning system regulates the development and use of land in the public interest, and has an important role to play in achieving sustainable waste management. It needs to be ready to deal with the challenges in the new waste strategy through national and regional planning guidance, policies contained in development plans, consideration of individual planning applications, and underpinning research and monitoring.

1 *Integrated Solid Waste Management: A Life Cycle Inventory*. White, Franke and Hindle, 1995. Blackie Academic and Professional. ISBN 0-7514-0046-7.

KEY PLAYERS

3.22   A number of key players are involved in developing waste related plans:

- *central Government*, and the *National Assembly* in Wales, have an important part to play by ensuring that adequate national planning policies are in place in order that sustainable waste management practices can be exercised at the regional and local levels in terms of the location and use of facilities

- *Regional Planning Bodies* in England will need to apply national policies as part of the process of drawing up regional planning guidance. In Wales there are no formal arrangements for regional planning, although there are voluntary groupings of local planning authorities for collaborative working on issues including waste management

- *Waste Planning Authorities* are responsible for ensuring that an adequate planning framework exists. They are required to prepare a waste development plan which has to take account of national and regional planning policy guidance. Waste Planning Authorities also have responsibility for determining planning applications for waste management facilities. In Greater London and the metropolitan areas, Waste Planning Authorities are the London borough councils and the metropolitan district councils. Outside Greater London and the metropolitan areas, Waste Planning Authorities comprise County Councils, National Park Authorities and the newly created unitary authorities. In Wales, the unitary authorities are responsible for local waste planning policy and for determining planning applications for waste management facilities. The Unitary Development Plans will incorporate waste policies which will take account of national planning policy guidance

- *the Environment Agency* is responsible for environmental regulation of developments under the Environment Act 1995. Whilst this is separate to the planning regime both may need to consider common issues. It is essential therefore that there is good, effective liaison between waste planning authorities and the Environment Agency. Planning Authorities should not seek to make judgements on pollution control matters that are the proper responsibility of the Environment Agency, and visa-versa. The Government and the National Assembly support the Agency's work to develop twin tracking of planning and pollution control applications, to ensure that applications are dealt with as speedily as possible, and that all the information required to make a reasoned decision is available

- *the waste management industry* has an important role to play in achieving sustainable waste management, by providing the full range of facilities required to deal with existing and projected waste streams and meeting goals set out in the waste strategy. Waste management companies should work closely with local authorities to develop more integrated waste management facilities

- *environmental organisations* (including voluntary and community groups, and not-for-profit organisations) aim particularly to secure improved protection of the environment and can bring useful views and advice to the planning process

NATIONAL PLANNING GUIDANCE

3.23 Planning policy guidance on waste management in England is set out in Planning Policy Guidance Note 10 *Planning and Waste Management* (PPG10). The guidance provides advice about how the land-use planning system should contribute to sustainable waste management through the provision of the required waste management facilities in England. It explains how this provision is regulated under the statutory planning and waste management systems. It also defines the roles of the various parties and emphasises the importance of liaison and consultation at all levels. Furthermore, it provides general advice for site selection, and matters which need to be taken into account when preparing waste development plans and considering planning applications for waste management facilities.

3.24 In Wales, guidance on waste is set out in *Planning Guidance (Wales) Planning Policy First Revision 1999*, which is supplemented by a series of Technical Advice Notes (TANs). A draft Technical Advice Note – *Planning, Pollution Control and Waste Management* – was issued in 1996, but a final version was not published. The National Assembly intends to prepare a revised Technical Advice Note for waste by the end of 2000.

REGIONAL PLANNING GUIDANCE

3.25 New arrangements in England for Regional Planning Guidance are being put in place through a revision of Planning Policy Guidance Note 11 *Regional Planning Guidance* (PPG11). This proposes that Regional Planning Guidance will be prepared by Regional Planning Bodies, in collaboration with Government Offices for the Regions and other organisations. This is of particular significance in planning for waste management since Waste Planning Authorities cannot consider the needs of their own areas in isolation. Waste management solutions may sometimes cross planning areas, as well as regional boundaries. In some circumstances, local options for the management of some types of waste may not be available.

3.26 The Government has recommended, therefore, in PPG10 the setting up of Regional Technical Advisory Bodies. These will advise the Regional Planning Bodies and provide specialist advice on options and strategies for dealing with the waste that will need to be managed within each region. The preferred option or strategy will be reflected in Regional Planning Guidance. It is important that appropriate preferred options should be incorporated into the Regional Planning Guidance at the earliest opportunity.

> **Regional self-sufficiency**
>
> In England, PPG10 makes clear the Government's view that most waste should be treated or disposed of within the region in which it is produced. Each region should provide for facilities with sufficient capacity to manage the quantity of waste that they expect to have to deal with in that area for at least ten years.

3.27 There are no formal arrangements in Wales for regional planning although there are voluntary groupings of local planning authorities for collaborative working on issues including waste management.

DEVELOPMENT PLANS

3.28   In London and the metropolitan areas, and in Wales, Part I of the Unitary Development Plan sets out the broad framework and overall land use planning strategy for waste management within the regional context. Part II of the Plan gives detailed expression to the policies by, for example, identifying sites or areas within which specific types of development may be acceptable or, if that is not possible, criteria against which the suitability of planning applications will be assessed.

3.29   Outside of these areas the development plan for waste generally comprises the Structure Plan and the Waste Local Plan. The functions of these two plans are analogous to Parts I and II of the Unitary Development Plan. In preparing development plans Waste Planning Authorities have to take account of government policies as set out in Planning Policy Guidance notes and Regional Planning Guidance.

3.30   It is important that provisions in development plans should reflect the new waste strategy and Regional Planning Guidance as soon as possible, in practice at the early stages of plan preparation or when a plan is subject to review.

3.31   Whilst provision for dealing with waste streams is widely included in waste development plans, these have less commonly considered the implications for local waste management of major proposals for development such as housing or commercial centres. There is a need to consider the broader context of how waste might be collected efficiently and effectively and dealt with, as far as possible, nearby. In the case of major housing development, for instance, the feasibility of community heating schemes might be considered. This could also bring other local environmental benefits such as a reduction in traffic carrying wastes.

Getting the public involved in the planning process

DETR has published a Code of Practice on the preparation of various waste and minerals development plans, called *Local Plans and Unitary Development Plans – a Guide to Procedures*. The document (product code 99PD0724) is available from the DETR Free Literature Unit at Wetherby (PO Box 236, Wetherby L23 7NB, telephone 0870 1226 236 or fax 0870 1226 237). It is designed as a simple guide to help the public take part in local plan inquiries, and sets down the procedures from plan deposit to adoption.

In Wales, guidance on the procedures for the preparation of unitary development plans is provided in *Planning Guidance (Wales) Unitary Development Plans 1996*, which sets out the requirements for publicity and consultation to ensure the local people and interested bodies are fully involved in making decisions on the policies and development proposals in their area. Further information is provided in the booklet *Development Plans: what you need to know* published by the former Welsh Office in 1997 which includes a revised Code of Practice from the deposit of a development plan through to its adoption by the local planning authority. The *Town and Country Planning (General Development Procedure) Order 1995* sets out the procedures connected with planning applications and appeals including the requirements for publicising applications for planning permission through site notices, serving notices on adjoining owners or occupiers and local advertisement, for consultations to be undertaken before the grant of permission and for representations to be taken fully into account by local planning authorities in determining planning applications for planning permission.

LGA Publications has issued the Planning Officers Society's good practice guide *Public Involvement in the Development Control Process*. Details of all LGA publications, and information on how to obtain them, can be found on the LGA website at www.lga.gov.uk, or alternatively telephone 020 7664 3131.

PLANNING APPLICATIONS

3.32   Since waste management facilities are seldom popular with those who live near proposed sites, it is important that waste management companies should discuss proposals with Waste Planning Authorities at an early stage, and should explain proposals thoroughly to local interests and to community groups. Careful attention should be paid to making proposals as environmentally and socially acceptable, as well as economically viable. Careful liaison is also needed with the local Planning Authority where this is not also the Waste Planning Authority, and with the Environment Agency to make sure that problems are addressed and that planning and licence conditions are complementary.

KEY ACTIONS FOR EFFECTIVE WASTE PLANNING

3.33   The drive to a more sustainable waste management system, with its lessening reliance on landfill sites, means that there will be a greater need for waste sorting and bulking depots, and materials and energy recovery facilities, in the future.

3.34   Waste management facilities are not popular neighbours. Thus it is essential that the preparation and adoption of waste development plans and their subsequent implementation through the approval of planning applications, is both open and accessible. The Government and the National Assembly are keen to see:

- all relevant organisations implement fully, and as soon as practicable both: the planning guidance contained in PPG10 and PPG11 in England; and in Wales the Planning Guidance (Wales) Planning Policy First Revision 1999 and, when finalised, in the Technical Advice Note for Waste

- all Waste Planning Authorities and developers thinking more holistically about the need to provide waste management facilities in line with this strategy, including consideration of how facilities can be better co-ordinated in relation to major new developments of, for instance, housing and commercial centres

- Waste Planning Authorities and the Environment Agency work closely on all planning applications, to ensure that adequate information is available to enable proper decisions to be made, and that planning and licence conditions are complementary and effective

- the waste management industry acting more proactively to developments in waste policy and waste planning, and in particular thinking more innovatively about new waste management facilities; and to operate these facilities to high standards

- waste companies consulting the Waste Planning Authority, Environment Agency and especially the local community – businesses and householders – about proposed waste management facilities at the earliest opportunity, in order to better deal with local concerns

Waste regulation and licensing

3.35   A waste management licence is required by anyone who proposes to deposit, recover or dispose of waste. Licences are issued by the Environment Agency in England and Wales.

3.36   The waste management licensing system under Part II of the Environmental Protection Act 1990 was brought into force for most sectors of industry in May 1994 and extended to the scrap metal industry in April 1995. Before being brought into force the provisions of Part II of the 1990 Act were modified to ensure fulfilment of the requirements of the amended Framework Directive on waste (91/156/EEC). The Waste Management Licensing Regulations 1994 make the necessary modifications and set out much of the detail of the licensing system. A key feature of the Directive, and the licensing system, is the objective of ensuring *that waste is recovered or disposed of without endangering human health and without using processes or methods which could harm the environment.*

3.37   The licensing system is separate from, but complementary to, the land use planning system. The planning system addresses the acceptability of a proposed development in terms of the use of the land. The purpose of a licence and the conditions attached to it is to ensure that the waste operation which it authorises is carried out in a way which protects the environment and human health. In line with the *polluter pays* principle, charges are made for the Agency's consideration of licence applications and an annual subsistence charge is payable by licence holders.

3.38   The Environment Agency must be satisfied that a licence applicant is a "fit and proper person" to hold a licence and three criteria are used to make this determination:

- the applicant, or another person involved in his business, has been convicted of an environmental offence. The Environment Agency has some discretion on this criterion and may take into account factors such the type of person convicted (for example an individual or a body corporate), the number of offences committed and the nature and gravity of the offences

- the management of the activities which are the subject of the application will be in the hands of a technically competent person

- the person to whom the licence is to be issued has made financial provision adequate to discharge the obligations of the licence

3.39   In most cases technical competence is demonstrated by the award of a Certificate of Technical Competence by the Waste Management Industry Training and Advisory Board (WAMITAB). For those waste facilities which are not covered by a WAMITAB certificate the assessment of technical competence is made by the Environment Agency. On the introduction of the licensing system, transitional arrangements were made for those people who applied for a Certificate of Technical Competence before 10 August 1994; these people were given a 5 year period in which to obtain the required qualifications. Most of the transitional arrangements ended on 10 August 1999 although a few exemptions will continue until 2004.

3.40   Once a licence is issued the Environment Agency is required to carry out *appropriate periodic inspections* of the site and to take the steps needed to ensure that the environment and human health are protected and that the conditions of the licence are complied with. The Agency has powers to suspend or revoke licences. On application, a licence may be transferred to another person if the Agency is satisfied that the proposed licence holder is a "fit and proper person". An application has to be made to surrender a site licence and the Agency may accept its surrender only if it is satisfied that the condition of the land, resulting from its use as a waste operation, is unlikely to cause environmental pollution or harm to human health.

3.41   The 1994 Regulations provide a range of exemptions from waste management licensing. Their main purpose is to encourage the recovery of waste. In each case the activity is exempt only if it is done in a way which complies with the terms of the exemption and does not endanger human health or harm the environment. In most cases, the exemption must be registered with the Environment Agency. The Environment Agency has a duty to carry out appropriate periodic inspections of exempt activities.

3.42   The Government is currently reviewing a limited number of the exemptions and we expect to publish a consultation paper in 2000.

3.43   There are severe penalties for recovering or disposing of waste without a waste management licence, in breach of licence conditions or the terms of an exemption, or in a way likely to cause environmental pollution or harm to human health. For example, on conviction in a Crown Court, imprisonment for 2 years and/or an unlimited fine.

3.44   The Government has asked the Environment Agency to continue a vigorous policy of prosecuting, where it has the evidence, anyone who illegally disposes of waste on sites which have neither a waste management licence nor a licensing exemption registered with the Environment Agency.

3.45   The waste management licensing system is supplemented by requirements for persons transporting waste in the course of their business (waste carriers) or acting as waste brokers to be registered with the Environment Agency.

Duty of Care

3.46   The Duty of Care applies to anyone who imports, produces, carries, keeps, treats or disposes of waste or, as a broker, has control of it. Everyone subject to the duty of care must take all such measures as are reasonable in the circumstances to:

- prevent contravention by any other person of the waste management provisions of the 1990 Act

- prevent the escape of the waste from his control or that of any other person

- ensure that waste is transferred only to an authorised person, such as the holder of a waste management licence, a person operating under the terms of a licensing exemption registered with the Environment Agency or a registered waste carrier

- ensure that a written description is transferred with the waste

3.47   Everyone to whom the Duty of Care applies has a legal obligation to comply with it and there are severe penalties for failing to do so. For example, an unlimited fine on conviction in a Crown Court. Practical guidance on complying with the Duty of Care is provided in *Waste Management, the Duty of Care, A Code of Practice*. The Duty does not apply to householders and household waste produced in the home.

Waste and integrated pollution prevention and control

3.48   The Integrated Pollution Prevention and Control (IPPC) Directive (96/61/EC) requires Member States to prevent or, where that is not possible, to reduce pollution from a range of industrial and other installations, by means of integrated permitting processes based on the application of *best available techniques*. The integrated approach takes a wide range of environmental impacts into account. These include emissions of pollutants to air, water and land; energy efficiency; consumption of raw materials; noise and site restoration. The aim is to achieve a high level of protection for the environment as a whole. Permits must take into account local environmental conditions at the site concerned; its technical characteristics and its geographical location.

3.49   The IPPC Directive applies to the following operations which are presently subject to control under Part II of the 1990 Act and the waste management licensing system:

- installations for the disposal or recovery of hazardous waste[2] with a capacity exceeding 10 tonnes per day

- installations for the incineration of municipal waste with a capacity exceeding 3 tonnes per hour

- installations for the treatment of non-hazardous waste[3] with a capacity exceeding 50 tonnes per day

- landfills receiving more than 10 tonnes per day with a total capacity exceeding 25,000 tonnes, excluding landfills of inert waste

3.50   The transposition of the IPPC Directive has been the subject of a series of consultation papers. It is proposed that key features of waste management licensing should be retained in the regulation of those waste management installations subject to the IPPC Directive. For example, the test of "fit and proper person" and appropriate provisions for the surrender of permits.

2 R1, R5, R6, R8 and R9 of Annex IIB to the Framework Directive on waste
3 D8 and D9 Of Annex IIA to the Framework Directive on waste

CHAPTER 4

Waste stakeholders

4.1   Working towards sustainable waste management requires the commitment of all the different groups of waste producers in society, in co-operation with the authorities and businesses responsible for regulating and controlling waste management.

4.2   The following diagram of the waste management cycle gives a simple overview of who generates waste, and how that waste is dealt with. A vast quantity of waste is generated in England and Wales – around 400 million tonnes every year. 27 million tonnes of this waste comes from households, with a further 48 million tonnes generated by manufacturers and 30 million tonnes by other businesses.

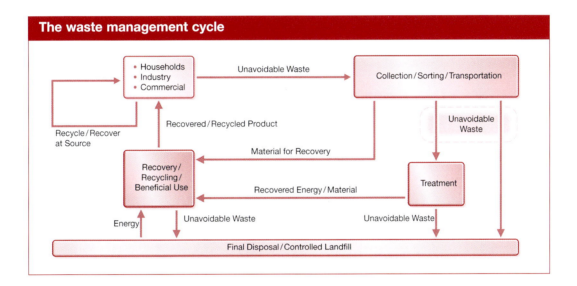

4.3   This chapter focuses on the roles and responsibilities of each sector of society which generates and manages waste. Waste generators (such as businesses, consumers and various Government operations) are considered in the first part of the chapter. This is followed by a closer look at how waste is regulated by the Environment Agency, and at those organisations responsible for collecting, managing and disposing of that waste.

4.4   Definitions for the various categories of waste are given in Chapter 2 section 2.2 of this part of the strategy.

Identifying the waste generators

4.5   Waste is generated both in the workplace and in the home. This section looks more closely at what businesses and individuals can do both to reduce the amount of waste they generate, and to deal effectively with the waste that is generated. It also considers waste management across Government, including the National Health Service and the Ministry of Defence Armed Services.

BUSINESSES

4.6 Waste can be a cost to business which is not always recognised. The true cost of waste is not just the cost of disposing of the waste a company produces, but also the costs of materials, energy, resources and staff input that went into the materials that become waste.

4.7 Waste can also be an opportunity for business, with far-reaching environmental and financial benefits accruing from effective waste reduction in the manufacturing process, re-use of refurbished or rescued components, and use of recycled or secondary materials in the manufacture, transportation and use of products. For example:

- recycling metals reduces impacts on the natural environment that would otherwise come from the quarrying, transportation, smelting and processing of virgin ore

- composting, or recovering energy from, organic waste like food and paper can directly reduce the methane emissions normally produced when such waste is landfilled. It also reduces pressure on natural peat lands, and reduces the greenhouse gas emissions that arise from the exploitation of peat

- recovering energy from waste can reduce the need to burn fossil fuels, and contributes to the Government's renewable energy programme

4.8 Both business and Government have increasingly begun to take advantage of the opportunities offered by waste. Since 1994 the Environmental Technology Best Practice Programme has worked together with business to test out good practice and help apply opportunities for savings through waste reduction.

4.9 DETR is developing guidelines to help businesses measure and report on waste, which will be published in 2000. The guidelines will provide step by step advice to help businesses identify and measure key waste streams, to set measurable targets for improvement, and to publish the information as part of an environmental report. They will also provide best practice tips on how to reduce and manage waste, and include case studies from leading businesses. The business waste guidelines will form part of a set of guidelines to help businesses report publicly on environmental performance. Guidelines that have already been developed can be viewed on the DETR website at www.environment.detr.gov.uk/envrp/index.htm .

Waste reduction in businesses

4.10 Waste production can be a result of inefficiencies in production or management processes. Many companies in England and Wales may lack the knowledge to tackle the unnecessary waste their activities produce, and may be unaware that they are producing more waste than they need to.

4.11 There is one very good reason why companies should be keen to reduce the amount of waste they generate: profit. Companies may not identify the full cost of the waste they produce. Waste disposal may often be a substantial cost for a business. Although these are issues of management time and funds spent on waste reduction, reducing waste can improve profit. Furthermore, the feed-through to lower input requirements can be important; raw materials are often 30 – 60% of turnover.

Business and the Environmental Technology Best Practice Programme

4.12   Many companies who have adopted waste reduction initiatives, often as participants in a waste minimisation club, have achieved savings which they did not initially expect. The Environmental Technology Best Practice Programme (ETBPP) aims to demonstrate the benefits of managing resource use and reducing environmental impact to companies across the whole of the UK, building on the lessons of the first waste minimisation clubs.

4.13   Information is produced for specific industry sectors such as chemicals, paper and board, textiles, and food and drink manufacture and retail. Guidance also covers generic themes of waste reduction and cleaner technology. This is published in the form of benchmarking guides which enable companies to assess their environmental performance; case studies demonstrating good practice; and guides which show how to implement environmental technologies and techniques.

The North Derbyshire Waste Management Group

Twenty two small and medium sized companies in Chesterfield and the Holmewood Industrial Estate near Clay Cross have joined together to form the North Derbyshire Waste Management Group. As part of their activities, the Group has organised two parallel series of five breakfast workshops. These are broadly based on established workshops developed by ETBPP. The early morning sessions look at:

- an introduction to waste reduction
- gaining commitment and getting started
- process mapping
- energy management
- tracking water use, packaging reduction or solvent management

Of necessity, the sessions are shorter and more focussed than the ETBPP half day workshops, but still provide ample time for companies with similar environmental challenges to discuss important issues across the breakfast table, before and after the delivery of key information.

4.14   When the ETBPP started in 1994, there were three waste minimisation clubs: Aire and Calder, Project Catalyst and Leicester. The Programme encourages the establishment of new clubs that are self-sustaining, with members working to help themselves. The clubs offer a good vehicle for engaging business in making environmental improvements, and can form a good route for disseminating information from the ETBPP.

4.15   Clubs have involved a variety of supporting partners, including Local Authorities, the Environment Agency, Business Links in England, Business Connect in Wales, and water companies. Clubs have gained income from bodies such as these as well as other sources such as the Welsh Development Agency, Local Challenge and European development budgets. It is estimated that around 100 self-sustaining clubs would make good coverage across the UK. Around 70 have now been established, and at least a further 15 or so are proposed. The ETBPP is continuing to provide advice to potential new clubs, and also trains club participants to hold their own workshops using ETBPP material.

ETBPP Good Practice Guidance

The Environmental Technology Best Practice Programme regularly publishes:

- benchmarking guides which enable companies to assess their environmental performance
- case studies demonstrating good practice
- guides which show how to implement environmental technologies and techniques

The ETBPP also addresses good management practice and raising awareness of the impact of waste on companies within management and staff. All the publications emphasise the practical results of introducing good practice measures, and in particular showcase the cost and resource savings that can be made from introducing more effective waste management practices.

The ETBPP's publications are disseminated through a wide range of channels, including trade associations, Business Links in England, Business Connect in Wales, the Environment Agency and the Wales Environment Centre, which gives detailed advice to business on waste reduction issues.

Information on the current publications list, together with recent publications are available over the internet at www.etbpp.gov.uk and also at www.etsu.com/etbppnews

---

Encouraging action by business on waste

4.16   The primary responsibility for commercial and industrial waste lies with the businesses that produce that waste. Many industry sectors already have an established waste re-use or recycling sub-sector. Companies may not even realise that by selling on waste materials, or using them in a different industrial process, they are in fact re-using and recycling waste.

Schroders

Schroders is a leading investment banking and asset management group. The company has 6,200 employees and offices in 36 countries and territories. Schroders (London Group) has taken a company-wide, strategic approach to identifying their main waste related environmental impacts and setting priorities for improvement. Waste management (in particular waste to landfill) has been identified as a priority action area in its newly revised environmental policy. The company believes that an environmentally responsible business is a well managed business and is committed to adopting best practice in line with Government guidance.

Effective communication is fundamental to Schroders's work and to its environmental improvement plan. The use of technology such as e-mail to communicate both internally and with clients has improved business efficiency and reduced demand for paper. Staff information and notices are now communicated electronically through the organisation's intranet, saving around 9,000 sheets of A4 paper each week.

4.17   A number of regulatory and non-regulatory initiatives have been launched over the past few years, the direct or indirect purpose of which has been to make business take its waste more seriously.

4.18   The *Landfill Tax* provides a strong economic signal to reduce reliance on landfill. Evidence suggests that the introduction of the tax has increased business interest in recycling and the use of inert waste in construction projects. The standard rate of landfill tax is to increase from the 1999 rate of £10 per tonne, by £1 each year to £15 per tonne of active waste by 2004. This will increase the financial incentive for companies to reduce the amount of waste they produce at source, to keep inert and active wastes separate, to seek out markets and buyers for waste (such as paper, food or plastic) that can be recycled or composted, and to consider the possibility of recovering energy from wastes with high calorific values.

4.19   Regulations to ensure *companies meet basic standards of good practice*. Regulation of
        emissions to the atmosphere and water, and for the safe handling and disposal of hazardous
        wastes, requires business to manage and dispose of the waste they produce safely and with
        minimum impact on the environment. Good regulation can be effective in encouraging
        some movement towards waste reduction, re-use or recovery by businesses. However, the
        main strength of regulation is to ensure minimum standards are maintained or raised.

4.20   Encouragement of *sectoral producer responsibility initiatives*, in which industry takes greater
        responsibility for their products when these have become waste. Such initiatives can
        encourage industry, commerce, communities and householders to cooperate to increase the
        re-use, recycling and recovery of materials. Voluntary producer responsibility initiatives
        have included nickel-cadmium batteries, and tyres. On packaging more formal, legally
        binding Regulations were introduced in 1997 to achieve targeted increases in recovery and
        recycling of packaging waste and to encourage greater re-use of packaging. With effect
        from 2000, retailers will also have an obligation to inform customers about a range of
        recycling matters (see Chapter 7 of this part of the strategy for more details on the
        packaging Regulations).

4.21   Provision of *information to business on best practice*, helping companies benchmark their
        performance in energy and resource use, transferring information and know-how across
        business sectors, and overcoming the information barriers to the adoption of best practice.
        A programme to promote sustainable waste management is being established (as detailed
        in Chapter 2 section 2.22 of this part of the strategy), with an aim of providing this advice
        to businesses and others. The Environmental Technology Best Practice Programme will
        also continue to offer advice and information to businesses, in particular through the
        Environment and Energy Helpline. Additionally, the Environment Agency is able to
        provide advice to companies (and others) on best practice and waste management
        reduction and recovery.

4.22   The purpose of *sectoral sustainability strategies* is to provide the framework for addressing the
        "triple bottom line" of economic, environmental and social aspects. Dealing with waste
        and the issues it raises for a given sector may be more actively addressed by businesses
        acting in concert, for example in improving the recyclability of products and in organising
        the recovery and recycling of wastes from the production process.

4.23   The Government is encouraging trade associations and other representative bodies to
        develop and implement strategies within at least six sectors by the end of 2000. The more
        pro-active sectors to date include: the Aluminium Federation, the British Plastics
        Federation, the British Printing Industry Federation, the British Retail Consortium, the
        Chemical Industries Association and the Society of Motor Manufacturers and Traders.
        The Government has also begun discussions with the British Cement Association.

Helping you to Club together

The Environmental Technology Best Practice Programme provides guidance to local organisations throughout the UK on setting up waste minimisation clubs. The map shows the location of existing or proposed clubs known to the Programme.

The Programme can help:

- by giving advice to local organisations and trade associations on running workshops;
- by supplying free publications;
- in some cases by providing free, short site visits to companies;
- by answering companies' queries through the Environment and Energy Helpline on 0800 585794.

On-going Waste Minimisation Club · **Club where there is no current activity** · Proposed Waste Minimisation Club

1. Highlands
2. Grampian
3. Tayside
4. Tayside (Food sector)
5. Fife
6. East of Scotland
7. North Glasgow
8. Clyde
9. Ayrshire (Textiles) Group
10. Northumbria
11. Project Tyneside
12. Carlisle
13. Sunderland
14. Easington
15. West Cumbria
16. Sedgefield
17. Teesside
18. WEFT (Textiles)
19. Belfast
20. Armagh & Banbridge
21. Barrow
22. Morecambe Bay
23. Selby
24. Keighley
25. Bradford
26. ELBEN
27. Aire & Calder
28. Humber
29. Project Catalyst
30. Bury
31. Knowsley
32. Merseyside
33. Environet 2000
34. Stockport
35. North Wales Waste Network
36. Dee
37. Huyton
38. North & Mid Cheshire
39. Don, Rother & Dearne
40. Lincoln (Food sector)
41. Lincolnshire
42. North Kesteven
43. Nottingham
44. Stoke (Ceramics)
45. Staffordshire Trent Valley

46. Shropshire
47. Black Country
48. West Midlands
49. Leicester Phase II
50. East Anglia (Food sector)
51. Corby
52. The Marches
53. Stratford
54. Northamptonshire
55. Worcester
56. Hereford
57. Bedivel
58. Suffolk
59. Essex
60. Caerphilly
61. SABINA
62. South Gloucester
63. Midsomer Norton
64. West Wiltshire
65. Vale of White Horse
66. Thames Valley
67. Slough
68. Surrey
69. Medway & Swale
70. North Devon #2
71. Somerset
72. Southampton
73. South Wessex
74. East Devon
75. Plymouth #3
76. East Cornwall
77. South Lakes
78. Penrith
79. Isle of Wight

*CLUBS IN LONDON
80. Camden
81. North & East London
82. PREMIER
83. South Thames
84. Harrow
85. Merton
86. Bromley & Bexley

87. North Derbyshire
88. Hereford & Worcester
89. Bath
90. Wigan
91. Oldham
92. Rochdale
93. Crieff & West Strathearn
94. Dumfries
95. Aylesbury Vale
96. Cardiff Bay
97. South Devon
98. West Cornwall
99. Swindon
100. South Bishop Auckland
101. County Durham

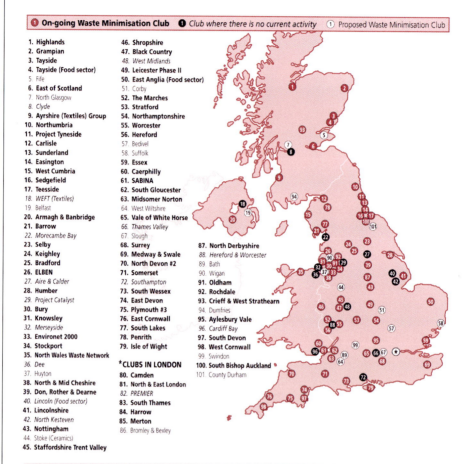

For further information on the free services available from the Environmental Technology Best Practice Programme please phone the

Environment and Energy Helpline 0800 585794

world wide web: **http://www.etbpp.gov.uk**   e-mail: **etbppenvhelp@aeat.co.uk**

The Environmental Technology Best Practice Programme is a Government programme. It promotes the use of better environmental practices that reduce costs for UK industry and commerce.

Source: DTI

Dee catchment waste minimisation project

The Dee waste minimisation project involved 13 companies actively participating in waste reduction at source. These companies were from a wide range of industries in the River Dee catchment area (from Snowdonia to Chester). The project was initiated by the centre for the Exploitation of Science and Technology, supported by the BOC Foundation, the Environment Agency and the former Welsh Office. The project, which started in 1995, ran for approximately 18 months during which time the participating companies adopted programmes to reduce waste generation at their sites.

All the participating companies identified opportunities for financial savings and for simultaneously reducing emissions and discharges to one or more of the three environmental media. The Project achieved financial savings amounting to £4.55 million per annum with an additional £1.2 million of potential savings identified. Most of these savings involved zero or low implementation costs and a payback of less than 1 year.

The identified opportunities represented:

- a total potential reduction of 130,000 tonnes per annum to landfill

- a reduction of water consumption of 600,000 cubic metres (with a similar reduction in the volume of liquid effluent)

- energy savings of 35,000 MWh – this was mainly electricity and was equivalent to a reduction of 25,200 tonnes per annum of CO_2 emissions to the atmosphere

- a reduction of 377 tonnes per annum of other emissions to air

Of the total number of opportunities identified, 45% involved technology modifications, 38% were procedural changes and 13% involved recycling or re-using material.

Only three of the participants did not experience significant changes in personnel, organisational structure or ownership during the course of the project. This reflects the changes that are now common throughout industry but also led to some barriers to implementing and maintaining the waste reduction programme. However, these barriers were overcome and almost all of the participants believed that the project met their expectations.

Practically all of the companies believed they would continue with their waste reduction activities after the project finished, For most companies this will be assisted by the implementation of an environmental management system (EMS). By the end of the project four companies had an EMS in place, three were installing a system and the remainder intended to do so sometime in the future.

A list of waste initiatives for business action

4.24   The range of initiatives which businesses can undertake on waste reduction and sustainable waste management is growing as companies and consumers become more aware of how, where, when and why waste is generated. A number of these ideas come from manufacturers and retailers themselves, either as part of the manufacturing vision or marketing strategy.

4.25   Other initiatives have been championed by consumer groups, community groups or environmental campaigners. But wherever the ideas arise from, they cannot be turned into reality without commitment from businesses to act on them, in partnership with others wherever possible.

4.26   *Ecodesign* is the process of producing more goods with fewer resources and less pollution, redesigning and re-manufacturing goods and services to enable recycling, or reducing harmful effects when they are returned to the environment. Producing more with less entails innovating in the way raw materials are extracted from the physical environment and used in the production process. This concept has been used by many companies to develop new working methods resulting in benefits both for their business and the environment.

4.27 **Consumer marketing and information programmes.** Retailers are acting to improve the information flow to their customers, so that consumer choices about purchase, use and disposal are better informed about key issues. Product information for consumers is one of the tools that can enable change to happen. By identifying issues for the public, and allowing products to be compared on the basis of environmental performance, it lays a foundation for further action.

4.28 **Product labelling.** Many companies already label products that are made from secondary materials or are recyclable. Environmental information on products can have an important role, particularly when used in conjunction with other market measures. The Government is keen to see this information provided to a proper standard. The *Green Claims Code* published in 1998 set out basic principles to help businesses make environmental claims about their products, including claims about their recycled content or recyclability.

4.29 More detailed guidance for businesses is now available in the form of a new international standard ISO 14021. The standard explains good and bad practice in making green claims and provides detailed information on claims, symbols, descriptions and verifications. The *Green Claims Code* is being updated to form a user-friendly introduction to it. The Government and the National Assembly want to encourage businesses to make full use of the new ISO standard and the Code, for reasons of competitiveness and good trading practice, as well as the positive effects of a better informed market for good producers.

4.30 **Environmental management systems.** Widening the scope of the management system to embrace the many other aspects – including energy and water consumption, emissions, raw material use and product stewardship – provides further opportunities to secure savings and reduce the environmental impacts of business. The international standard for environmental management systems, ISO 14001, and the EU Eco-Management and Audit Scheme (EMAS) are increasingly being seen as valuable tools. This is reflected in their increasing use by business to secure savings, enhance efficiency and demonstrate the commitment of business to manage its impacts and continuously improve its environmental performance.

4.31 **Making a Corporate Commitment (MACC).** Whether through the management systems approach or not, the Government is keen to see more businesses measuring, monitoring and setting targets to improve performance in specific areas. To this end it intends to relaunch MACC to focus on resource efficiency. The commitment is to make a public declaration to meet quantified targets within a set period. One of the key areas for demonstrating improvement in resource utilisation is waste. Improved performance in waste is clearly important for any business both for profit efficiency and cost reduction, but it is the environmental benefits that increasingly capture the attention of employees and external stakeholders. MACC provides another practical tool to encourage and maintain commitment to all levels in the organisation to reduce waste. Target setting is central to MACC. But by additionally requiring participants to report on progress against their targets, MACC also links with the drive for wider environmental reporting.

4.32 **Business reporting.** Increasing numbers of major companies are reporting publicly on their environmental performance. The most forward looking companies have got well beyond treating this as a public relations exercise (or *greenwash*). They are identifying their major environmental impacts, setting targets for improvement and monitoring performance. Public commitment to improvement is part of motivating real change. Some of the more innovative companies are reporting not just on the waste they produce, and their targets for waste reduction, but on initiatives to redesign production and product management to create opportunities for using recycled materials.

4.33   Recent surveys have shown levels of environmental reporting are high in the top utility companies, where 94% of top 350 companies produce a separate environmental report, and in minerals/resource companies where almost half do so. But environmental reporting is still infrequent among other sectors – general industries, consumer goods and retail for instance. All major companies are encouraged to report publicly on their environmental performance, and their reporting should address key environmental impacts, including impacts from waste.

4.34   Too few companies set themselves targets for reducing environmental damage and gaining added value from key waste streams. Only 45 companies in the top 350 provide quantified measures of waste products, or of amounts recycled or sent for energy recovery treatment as an alternative to disposal. Again, such reporting tends to be by utilities and resource companies, with too few companies reporting from other sectors.

4.35   *Producer responsibility.* If recycling levels for a product or material are low, reflecting market failure, then the Government, or in Wales the National Assembly, will consider whether a producer responsibility approach is likely to be a cost effective way of increasing recovery and recycling levels. Selection of a product or material sector for producer responsibility treatment will be dependent upon an assessment of its end-of-life negative environmental impact (volume, weight or hazardous nature) set against the net costs which would be associated with actions by producers to reduce that impact by recovery or recycling.

4.36   A statutory approach will be considered where a voluntary solution would be unlikely to achieve the desired benefits and where those benefits are significant as against the likely costs of regulation; or where there is a need to demonstrate compliance with formal EU requirements. In such circumstances the Government and the National Assembly will consider using the enabling powers contained in the Environment Act 1995.

4.37   *Life cycle assessment* (Chapter 3 section 3.12 of this part of the strategy) can be a useful tool to help evaluate the environmental performance of a product *from cradle to grave*. Life cycle assessment identifies the materials, energy and waste streams of a product over its entire life cycle – through manufacture, use and disposal – so the environmental impacts can be identified. It can be used by businesses to help identify changes to their operations which can lead to significant cost savings and environmental benefits.

4.38   Life cycle assessment is also one of the principles behind the new more integrated approach to product policy. Integrated Product Policy looks at the relationship between products and their burden on the environment throughout their life cycle, selects the priority areas for improvement and decides on the policy and market measures best suited to deliver the improvement.

4.39   *Supply chain management.* Increasingly supply chain pressure is being exerted by business through contract specifications and procurement policies reflecting the need to reduce waste and cut costs, for example by minimising the handling of packaging waste. Equally in trying to meet other targets, producers are stimulating the development of new techniques which use less material – and hence give rise to less waste. So in order to achieve higher fuel efficiencies in their vehicles, motor manufacturers are specifying lighter body parts and engines, using thinner gauged or lighter metal and alloys that can be recycled at the end of a vehicle's life.

4.40   ***Small business best practice guidance.*** A significant amount of commercial and industrial waste comes from small and medium-sized enterprises – with fewer than 250 employees. Best Practice help is available for these companies, and the Government will implement its plans to provide an improved service through the Environment and Energy Helpline. However, many companies face difficulties in accessing the full range of disposal methods. The new Waste and Resources Action Programme will work with the waste management industry and local authorities to explore how the range of services available to smaller businesses – particularly for separate collection of recycled materials – can be extended and improved. It will also consider how best to provide guidance to smaller businesses on identifying the Best Practicable Environmental Options for their wastes.

HOUSEHOLDERS

4.41   Approximately 25 million tonnes of household waste is produced in England and Wales each year. The range of materials included in this waste is large. The indications are that household waste typically includes:

- 32% paper and card
- 21% putrescible wastes
- 9% glass
- 8% miscellaneous combustible wastes
- 7% fines
- 6% ferrous metals
- 6% dense plastics
- 5% plastic films
- 2% textiles
- 2% non-ferrous metals
- 2% miscellaneous non-combustible wastes

4.42   We all have a vital role to play in tackling waste. While industrial and commercial activity (which in itself is consumer driven) produces far more waste then we – as individual householders – produce, managing our household waste costs local councils a significant amount of money. Net local expenditure by English local authorities in 1998/99 on waste management was £1.3 billion, with a further £73.7[1] million spent by Welsh local authorities in 1998/99.

4.43   There are a number of problems associated with getting people to manage their domestic waste more sustainably. *Lack of information* – a lack of knowledge about where, when, why and how much waste people generate in their own homes represents a real barrier on the path towards sustainable waste management. The Government and the National Assembly for Wales are committed to tackle this problem, in close partnership with local government, organisations and businesses across the spectrum of society.

4.44   *Denial of responsibility* – people tend to believe that putting their waste in the rubbish bin is the end of the story. In fact it is the first step in a chain of actions (collection, transportation, sorting, storage, treatment, recovery, disposal) which we as a society must make more sustainable.

1 Budgeted expenditure estimate

4.45   *Denial of power* – even when people do worry about the amount of waste their household generates, there is a belief that they are powerless to stem the flow of waste through their home. This is not the case. Individuals do have power to affect the amount of domestic waste they produce, by choosing more durable products, or those with less packaging. Householders can also help by sorting their waste before it is collected. Finally, individuals can have an impact through using their voice as part of the decision-making process, which is looked at in more detail in the box *Getting the public involved in the planning process* in Chapter 3 of this part of the strategy.

4.46   *Lack of supporting infrastructure* – most people are willing to recycle their waste if there is an infrastructure available which makes waste separation and recycling simple for them to do. The lack of infrastructure (such as kerbside collection services and community composting facilities) is a real problem for much of our waste, and one which the Government and National Assembly are committed to tackling in partnership with local councils and businesses. But it can be overcome, and it does not always need Government action to put in place effective local waste recovery schemes, such as the old jumble collection services run by various charity shops in a number of local areas.

4.47   Raising waste awareness through publicity campaigns and availability of information is considered in Chapter 2 section 2.29 of this part of the strategy. The following sections concentrate on the actions we can all take to reduce the amount of waste we generate, and the actions we can take to manage our domestic waste more sustainably. (Details of the various household waste incentive schemes we are investigating can be found in Chapter 5 section 5.7 of this part of the strategy).

Household waste reduction

4.48   Up until now the majority of waste reduction initiatives have taken place within industry. However, recent developments mean that there is more scope for action by householders and consumers. Waste enters the home in a number of guises. Excessive packaging and junk mail are obvious examples of waste, which normally find their way straight from the front door into the rubbish bin. Other waste is less obvious – leftover food, or household appliances which are greedy for resources but produce more waste than end product. Some packaging is essential to keep products (such as food) safe, and only becomes waste when it has served its purpose. Similarly, some products (such as newspapers) only have a short lifespan and end up in the rubbish bin after a few days. Other products – furniture, interior structures, household appliances, clothes, etc – have a lifespan lasting years, but even these will one day become waste – often before they need to.

Junk mail

4.49   Junk mail can be a significant, and often unwelcome element of the household waste stream. The number of items sent to consumers in England and Wales doubled from 1.5 billion in 1990 to 3.28 billion in 1999 (see also Chapter 3 in Part 1 of this strategy).

Reducing the flow of Junk Mail

People can ask to have their names removed from the mailing lists that direct sales companies (and others) use to distribute their literature by writing to the Mailing Preference Service at the following address:

Freepost 22
London W1E 7EZ
Telephone: 0345 034 599

People can also write directly to their bank or other companies that they receive junk mail from, to request that they are taken off mailing lists or that excess promotional material is not included with their statements, and remember to tick the box provided on many forms to tell companies not to send further details of their products and services through the post.

Recycling household waste

4.50 The constituents of dustbin waste vary according to the area of population, the relative economic and social factors of that population, and the method of waste collection. In general, recycling is optimised where the waste is composed mainly of one material, although alternative technologies are being proposed which deal with more mixed waste streams. Recycling requires sorting materials to ensure that they are of a quality equal to virgin material. Wastes that can be recycled include:

- paper (newspapers and magazines) and card
- dense plastics
- textiles
- glass
- ferrous metals (iron and steel)
- non-ferrous metals (mostly aluminium)

4.51 In addition, the organic, biodegradable or putrescible waste (such as peelings and other kitchen waste) can be composted (see Chapter 5 section 5.46 of this part of the strategy).

4.52 Furthermore, extra effort should be made to ensure certain products are not thrown away with the general household rubbish. For example, we currently dispose of around 600 million consumer batteries each year, some of which contain hazardous materials (further details on batteries can be found in Chapter 8 section 8.24 of this part of the strategy). Old refrigerators and freezers may also contain some hazardous materials (such as ozone depleting substances), and a number of schemes for collecting old equipment are in operation in various parts of England and Wales. And the substantial rise in home ownership of computers and related equipment has also led to a rise in obsolete or outdated equipment being throw out alongside the rest of the household waste – yet the circuitry in this equipment can often be put to good use in other products (which do not require overly sophisticated circuitry), if they are recovered in time.

Newport Wastesavers Charitable Trust

Newport Wastesavers Charitable Trust is a good example of the work of the community sector. It comprises a charity and a not-for-profit operating company.

The charity recycles furniture and white goods. During 1999, it assisted over 2,000 families, referred by registered agencies as being in need. The company has established partnerships with Newport County Borough Council and Gwent Probation Service, and operates a *green box* kerbside collection of recyclable material from 20,000 households in Newport. The service has achieved a materials recovery rate of 15% from participating households. By 2003, all of Newport's 50,000 households will be serviced fortnightly. An office paper collection operation is also collecting from over 200 organisations throughout south east Wales.

Community involvement in the decision-making process

4.53 The impact of householders (often working together) to influence either a council or a major retailer can be very influential. There are many avenues in place through which people can get their voices heard – such as community groups, parish and district councils and national voluntary groups.

4.54 People have a right to take part in the planning process, both at the stage where waste development, management, recycling and composting plans are being drawn up by local councils; and also when planning applications for waste management facilities are made.

4.55 Details on the decision-making process, and how individuals and organisations can make their voices heard as local waste management options are being drawn up, can be found in the box *Getting the public involved in the planning process* in Chapter 3 of this part of the strategy.

GOVERNMENT

4.56 The Government and the National Assembly have a number of roles to play to establish more sustainable waste management across England and Wales. In addition to their role of encouraging voluntary action, introducing, monitoring and amending various legislative and economic instruments to encourage and enforce the safer and more sustainable management of the waste we all generate, they also have a role in raising waste awareness across all sectors of society, in encouraging the various stakeholder groups to talk (and listen) to each other with the aim of establishing waste management partnerships. They also have a responsibility to ensure the waste they generate themselves is managed as sustainably as possible as an example to others. The following section concentrates mainly on the last of those points.

Greening Government

4.57 The Government is committed to improving the environmental performance of its Departments. It is working through the Green Ministers Committee to ensure this happens, and has established the Parliamentary Environmental Audit Committee to audit the Government's performance against its targets.

4.58 The Green Ministers Committee has agreed a Greening Operations Model Framework and Improvement Programme which all Departments have used to draw up their own strategies for improving the environmental aspects of their operations, including waste management.

4.59   Green Ministers also set out a waste target for 2000/01 in their first annual report published in July 1999[2]. This first Government waste target is: *to aim to recover a minimum of 40% of total office waste, with at least 25% of that recovery coming from recycling or composting.* The report makes clear that this is an initial target and that Green Ministers will want to consider if it can subsequently be extended to non-office waste. The report also sets out the various actions that Departments have taken individually.

Green Guide for Buyers

DETR has issued a *Green Guide for Buyers* which is relevant for all buyers in Government Departments. In addition to covering policy and practice, dealing with for example UK and EC procurement rules, grounds for rejecting suppliers and the importance of reducing waste, the *Guide* contains checklists to help buyers specify goods and services that are environmentally preferable. It has been complimented by a note jointly issued by HM Treasury and DETR on environmental issues in procurement. This reinforces the message that buyers should take account of life cycle costs and quality when determining value for money.

4.60   In Wales, the National Assembly will set an example by implementing the principles of sustainable development in its internal operations. One of the proposals in the Assembly's draft Sustainable Development Scheme is an annual review of its green housekeeping plan, which includes policies on waste management. In the current plan, the Assembly aims to reduce the amount of waste it sends to landfill, to improve existing recycling schemes and to look for new recycling opportunities where possible.

4.61   DETR has produced guidance to help Departments improve their waste management. This stresses that the main priority for Departments should be to reduce waste by using resources more efficiently, encouraging re-use and improving recycling schemes.

4.62   The importance of efficient procurement practices in Departments has been recognised and further guidance has been issued by DETR to help them achieve that. Chapter 3 of Part 1 sets out proposals for piloting a scheme to require procurement of certain recycled products. A joint note with HM Treasury on *Environmental Issues in Purchasing* emphasises the need to take decisions on the basis of whole life costs, rather than initial prices alone, and that this includes factors such as disposal costs. As well as advising public sector purchasers to look to waste reduction and re-use, the guidance recommends that purchasers should not generally look for new items when refurbished parts or products can be used.

Action within the Environment Agency

The Environment Agency has adopted good practice in waste reduction throughout its organisation. It has taken the broad view that waste reduction (or more effective resource management) should cover all the resources used within the organisation and has targeted those with the greatest environmental impact for action. These include:

- *energy use* in transport and in buildings
- *water* – to deliver a national target the Agency cut its water consumption by 30.4% in 1998/99 on a baseline year of 1996/97
- *Construction aggregates* – the Agency uses secondary or recycled aggregates where practicable: last year 38% of the aggregates used by the Agency were taken from secondary or recycled sources

Overall, the Agency is adopting a formal environmental management system. Three pioneer sites have already achieved ISO 14001 and the remaining operating sites are programmed to achieve this over the next 3 years.

In 1999, the Agency became the first Government body to publish an environmental report of its own activities.

---

2 This report can be found on the internet at www.environment.detr.gov.uk/greening/ar9899 – free hard copies can also be obtained by writing to DETR, Zone 5/A1, Ashdown House, 123 Victoria Street, London SW1E 6DE. Product code 99EP0395.

Ministry of Defence and the UK Armed Forces

4.63   The Ministry of Defence (MOD) is undertaking a number of actions in support of the Greening Government initiative and as part of its corporate environmental management system:

- MOD is measuring the level of waste production within a number of different areas rather than across the whole Department

- Phase-out plans for MOD PCBs (polychlorinated biphenyls) have been developed. All PCB waste is to be removed from equipment and destroyed by 2000

- Fluorescent light tubes are being recycled using a contract set up by MOD's Disposal Sales Agency. All expired disposal contracts are considered for the scheme

- Tenders have been sought for a call-off contract for the disposal of all types of MOD special waste

4.64   More generally on waste management issues, the following actions are being actively pursued:

- MOD policy is to reduce the production of waste wherever possible. Line managers are responsible for investigating and implementing opportunities for waste reduction which do not compromise operational requirements

- All procurement decisions should include consideration of potential waste implications and should address: maximum recyclability; maximum use of products based on recycled materials; minimum use of unnecessary packaging; minimum pollutant emissions

- All MOD personnel involved in handling, storing, treating, transferring or disposing of waste must comply with the Duty of Care

- Commanding Officers and Heads of Establishment are responsible for ensuring compliance with legislation, regulations and statutory requirements for waste producers, carriers and disposers

- Units or establishments which have significant quantities of waste (or special waste) should have a designated waste manager

- The waste manager has an audit responsibility for Commanding Officers and Heads of Establishment, and satisfies certain regulatory and licensing requirements

- Waste manager responsibilities vary between individual sites and activities

- It is MOD policy to reduce and re-use packaging wherever practicable and within the constraints of operational effectiveness

- MOD is working on an overall waste reduction strategy. Specific initiatives relevant to waste reduction include the Energy Efficiency Campaign and a Munitions Research Package which aims to improve the disposability of Explosive Ordnance

Department of Health and the National Health Service

4.65   The National Health Service can play an important role in sustainable waste management. Since the removal of Crown Immunity under the NHS and Community Care Act 1990, and the implementation of the Environmental Protection Act 1990, the health service has had to become more aware of healthcare waste disposal issues.

4.66   Many NHS trusts contract out the disposal of healthcare waste to private sector waste management companies. Trusts are also increasingly aware that technology provides a range of methods for treating healthcare waste, and some are adopting waste management strategies that rely to some degree on these new methods.

4.67   If the NHS can reduce the amount of waste produced by NHS trusts, the benefit will be twofold: NHS waste management costs will be contained, and a reduction in environmental impact can be expected. In the NHS, as elsewhere, effective waste management depends on:

- adequate segregation of waste
- adequate waste awareness (often reinforced by staff training)
- collecting better data on waste production

4.68   Most healthcare wastes represent a low level of threat to the environment. Such threats that may materialise can be managed by established techniques. Sustainability is not undermined by any unusual characteristics in the majority of healthcare wastes.

4.69   The NHS can also contribute to sustainability by incorporating environmental design within its specification and procurement techniques. This usually involves:

- discussions between the NHS and manufacturers of medical products about the design and packaging of medical devices

- the use of new materials for medical products so as to make them less toxic in production, recovery and disposal, and more easily recyclable

4.70   Medical products must reflect and satisfy the demands of medical practice and avoid prejudicing clinical outcome. This done, opportunities should be taken to eliminate or reduce clinical waste. The process is iterative and reciprocal: alterations in procedure may change the product and its packaging.

4.71   The NHS has to discharge its management responsibilities in accordance with statutory requirements and good practice, while still providing value for money. The NHS Estates Agency provides technical guidance to this end. The latest in the series is *Health Technical Memorandum 2065 – Healthcare waste management – segregation of waste streams in clinical areas*, issued in October 1997, and available from the Agency.

Regulating and managing waste

4.72   Even when waste production has been reduced to a practicable minimum, there will still be millions of tonnes of waste produced across England and Wales each year. Much of this will have the potential to damage our environment and has to be managed in accordance with the principles of sustainable development. The management of waste is tightly controlled, mainly through the provisions of the Environmental Protection Act 1990, overseen by the Environment Agency.

4.73 The collection, treatment and disposal of much of the waste generated by businesses and householders is undertaken by the waste management industry, though the collection and management of municipal waste is the duty of local waste collection and waste disposal authorities.

Waste in ports and coastal waters

This strategy refers only to waste generated and managed on land. The following policy documents deal with waste in ports and coastal waters:

- *Port waste management planning: how to do it* – London DETR 1998, ISBN: 1-85112-068-8. Copies are available from local marine offices or direct from DETR Shipping Policy 3, Zone 4/12, Great Minster House, London SW1P 4DR (telephone 020 7944 3898), or the DETR website at www.detr.gov.uk

- *Merchant Shipping (Port Waste Reception Facilities) Regulations 1997* SI 1997 No 3018, available from The Stationery Office, ISBN 0-11-065335-1

4.74 The Government's sustainable development strategy, *A better quality of life*, makes it clear that a significant improvement in resource efficiency is needed over the next 20 years. This does not mean making ourselves worse off, or denying others the benefits we already enjoy. But it does mean we need to do things differently. Disposing of products at the end of their lives can often represent a missed opportunity. Many existing products can be re-used, or have value (in the form of materials and energy) recovered from them.

4.75 Putting a more sustainable waste management system into place will involve local government, the Environment Agency and the waste management industry. But the need to engage the local community – householders, charities and community groups – and local businesses in the processes of sustainable waste management should never be overlooked or disregarded.

4.76 The regulation framework within which waste has to be managed in England and Wales is set out in the Waste Management Licensing Regulations (Chapter 3 section 3.35 of this part of the strategy) and the Integrated Pollution Prevention and Control Regulations (Chapter 3 section 3.48 of this part of the strategy).

THE ENVIRONMENT AGENCY

4.77 The Environment Agency was established in April 1996 to regulate emissions of pollutants to air, land and water.

Regulating and monitoring of waste activities

4.78 The Agency's main role in the sustainable management of waste is through its regulatory activities to protect the environment and human health. Carrying out its duties in the licensing, monitoring and inspection of waste management facilities is the single most important contribution the Agency makes to sustainable waste management.

4.79 To improve targeting of its inspection resources, the Agency shortly intends to introduce a risk-based system for determining the frequencies at which it inspects waste management facilities (OPRA). OPRA takes account of pollution risk and operator performance, and will enable the Agency to direct its inspections towards higher risk sites and illegal activities.

The Agency's role in promoting waste reduction

4.80   The Environment Agency also works with businesses across England and Wales to cut the costs – both financial and environmental – associated with the production of waste. Under the new IPPC regime many of the sites regulated by the Agency have a duty to cut unnecessary waste.

4.81   The Agency has supported moves by companies to reduce the waste they generate, through activities such as waste minimisation clubs, and by working with the Environmental Technology Best Practice Programme (ETBPP) to ensure collaboration and coordination of a number of initiatives in waste reduction. Increasingly, the Agency is targeting its waste reduction activities at specific industrial sectors identified from its Waste Production Survey, and specific types of waste (such as hazardous wastes) to achieve the most effective use of its resources in delivering environmental benefit.

Data and information

4.82   The Agency is committed to providing assistance to local authorities and the waste management industry, through the provision of relevant and up to date information on the amounts of waste and the management of it. The results of the Waste Production Survey have provided for the first time reliable estimates of industrial and commercial waste, which have informed this waste strategy. The Agency's survey has established a clear relationship between the amount of waste produced and the size of the company.

4.83   Additionally, as part of its role in providing information, more detailed information from the survey together with supplementary work being undertaken by the Agency will be published in reports for each region (known as the Strategic Waste Management Assessment) to provide information to the Regional Technical Advisory Bodies in England. These reports will contain information on the wastes arising, the facilities available and some assessments (using the Agency's life cycle tool) of the costs and benefits of different mixes of waste management options for dealing with municipal waste.

4.84   Apart from providing the information described above, the Agency has embarked upon a programme to produce the maximum benefit from the Waste Production Survey. During 2000, the Agency will provide key information (known as the Wastes Database) from the survey on its website. This information will be in two parts:

● the amount of waste produced by the best, poorest and average performers in each industry sector – this will enable businesses to compare their waste production with those for similar companies, and is intended to encourage waste reduction

● the waste materials produced by each industry sector – this will form the basis of a national waste exchange and allow those operating waste exchanges or seeking to re-use waste materials will be able to identify those companies most likely to produce it

THE WASTE MANAGEMENT INDUSTRY

4.85   The Government and the National Assembly for Wales believes that the waste management industry will be a primary player in delivering its strategy for waste. There is a lot of common ground between our aims – meeting targets, and where appropriate dealing with waste higher up the hierarchy – and the industry's need to meet its customers' expectations. The industry may be expected to change and develop in a variety of ways, as highlighted below.

4.86   It will be increasingly important for the industry to offer a range of local services, rather than just landfill or just incineration, in order that their customers can benefit from a real choice. Satisfied customers will benefit the industry and flexibility will help to meet this strategy's goals and targets. It is likely that a number of larger more integrated companies will come to dominate the market, although there will always be a place for smaller, localised companies, or specialist facilities such as high temperature incinerators.

4.87   Waste companies will need to respond rapidly, and anticipate changes brought about by regulation: slow moving companies will be threatened by changes, whereas companies with foresight are likely to reap commercial rewards. An example is the reduction of biodegradable municipal waste to landfill required by the Landfill Directive. Industry should make good use of the time available to make the necessary planning and investment in alternative treatments for biodegradable waste.

4.88   Embracing new techniques and developments, such as anaerobic digestion and pyrolysis, will give the industry a commercial advantage in the future. It is important that there is original thought and innovation in the industry. Increased recovery and reduction at source may lead to smaller quantities of more complex wastes which need technical solutions. The Government and the National Assembly welcome the setting up of an Environmental Body (the Environmental Services Association Research Trust – ESART) for research and development in this area.

4.89   Improving public confidence in the industry will be important for the industry's development, and continued competitiveness, most obviously in the planning process. (For further information on the waste planning process, see Chapter 3 section 3.21 of this part of the strategy).

4.90   The waste management industry, with the help of DTI as the industry's sponsoring department, is making efforts to address the public's perception of waste issues and management. An industry led group, representing a wide range of interests, has begun to look at this issue with a view to making recommendations on best practice in building public confidence. This is seen as an important step towards achieving more sustainable waste management. This group is working closely with the Environmental Services Association Research Trust, and a research project funded from the Landfill Tax Credit Scheme is now underway.

4.91   Providing a high standard of training to employees will be important in the future, for an industry wanting to respond to rapid changes. Companies which fail to train their employees to the relevant level of technical competence will face action by the Environment Agency. The Government and the National Assembly welcome proposals to develop a national training organisation for waste and will work with the industry and the Environment Agency to devise and implement properly accredited training in waste reduction techniques and other sustainable waste management options. The industry is already looking to develop and revise the awards in time for re-accreditation in March 2002.

4.92   Improved environmental standards can benefit those companies that practise them in the future. Apart from raising public confidence, these companies will benefit from Environment Agency proposals to move to a more risk based assessment towards waste facilities, as opposed to their current cyclical approach. The Government and the National Assembly welcome this development, which will help towards its aim of more sustainable waste management.

4.93   In all these developments affecting the waste management industry, it will be important to strike a careful balance between supporting the industry and helping to realise the objectives of the strategy, and allowing market forces to apply their own pressures. Although regulation cannot be ruled out, involvement will generally be indirect, for example through encouragement and provision of information, fiscal measures, or incentives such as the Landfill Tax Credit Scheme.

4.94   The Landfill Tax Credit Scheme provides a valuable opportunity for the industry to address many of the broad issues mentioned above, such as the need for greater innovation, training and improving public confidence. Industry should continue to make the most of this fiscal incentive.

WASTE COLLECTION ACTIVITY

4.95   The key to successful materials (and energy) recovery from waste is to ensure the waste undergoing these processes is as *homogenous* as possible. Homogenous waste can be defined as waste composed mainly of a single material (such as paper waste, or brown glass). Conversely, *mixed* waste can be thought of as waste made up of a wide variety of materials. Unsorted household waste is a good example of *mixed* waste stream.

4.96   Before waste can be recycled, it needs to be sorted into its component materials. Manufacturing processes often produce large quantities of waste composed of a single material, which can be recovered fairly easily. In contrast, household waste often has to be delivered to purpose built waste sorting facilities (Materials Recycling Facilities) where it can be separated into its constituent waste streams. It is at this point that waste will be shredded, crushed or bulked up to make its subsequent transport, treatment or disposal simpler, safer and often cheaper.

4.97   The method used to collect waste can therefore have a large impact on the practicability (and profitability) of subsequent recycling operations. This section considers the roles and responsibilities of the main players involved in waste collection – including the statutory duties of local Waste Collection Authorities.

Community Recycling Network

The Community Recycling Network is the umbrella organisation for over 250 community groups, co-operatives and not-for-profit businesses in the community waste sector. By providing advice, training, information and practical initiatives it aims to promote community-based initiatives as the most effective way of tackling the UK's growing waste problem. CRN members work in partnership with local authorities and business to develop the best practice in all fields of sustainable waste management including waste reduction and materials recycling. The CRN has links with equivalent organisations in Europe.

CRN represents the community waste sector in the UK, though it has been encouraging the development of regional recycling networks to improve support in some locations. Currently CRN members offer a householder-separated kerbside collection of recycled materials to over 800,000 households in the UK. CRN seeks to empower households and communities to take responsibility for their waste streams, these being the most visible manifestation of the individual's impact on his or her environment.

4.98   Waste Collection Authorities have a duty to collect household waste except in certain prescribed circumstances. They also have a duty to collect commercial waste if requested to do so and may also collect industrial waste. Waste collected, other than that which the authority makes arrangements to recycle, must be delivered to the appropriate Waste Disposal Authority.

4.99   Since 1 April 2000, local authority waste collection services have been subject to the Best Value duty under the Local Government Act 1999. This requires authorities to deliver their services to clear standards – covering both cost and quality – by the most effective, efficient and economic means available in order to deliver continuous improvements in service provision. Similarly in Wales, the National Assembly has formally repealed Compulsory Competitive Tendering from 2 January 2000 (although the requirement had been subject to a moratorium in Wales since October 1994) and introduced the duty of Best Value from 1 April 2000.

Best Value performance indicators

These indicators have been set up by the Government and reflect the interest in the delivery of local services. The indicators are designed to enable comparisons to be made between the performance of different local authorities and within an authority over time. Local authorities will need to set targets against the indicators to reflect the continuous improvements in performance required under Best Value. To ensure that the indicators give a balanced view of performance the Government has sought to reflect five *dimensions* of performance in the indicators. These are:

- *strategic objectives* – why the service exists and what it seeks to achieve
- *cost/efficiency* – the resources committed to a service and the efficiency with which they are turned into outputs
- *service delivery outcomes* – how well the service is being operated in order to achieve the strategic objectives
- *quality* – the quality of the service delivered, explicitly reflecting users' experience of the services
- *fair access* – ease and equality of access to services

Waste management is a key service provided by local authorities and a number of Best Value indicators have been set for waste management services. These are:

- of the total tonnage of household waste arisings
    - percentage recycled
    - percentage composted
    - percentage used to recover heat, power and other energy sources
    - percentage landfilled
- weight of household waste collected, per head
- cost of waste collection per household
- cost of municipal waste disposal, per tonne
- number of collections missed, per 100,000 collections of household waste
- percentage of people expressing satisfaction with
    - recycling facilities
    - household waste collection
    - civic amenity sites
- percentage of population served by a kerbside collection of recylable waste, or within one kilometre of a recycling centre

4.100 Waste Collection Authorities also have a duty to prepare and publicise waste recycling plans under section 49 of the Environmental Protection Act 1990 taking account of DETR/Welsh Office guidance to local authorities. A recycling plan is an authority's statement of the arrangements made and proposed for recycling household and commercial waste. It should:

- include information about the kinds and quantities of controlled wastes which the Waste Collection Authority expects to collect or purchase during the period of the plan, and which it expects to deal with by recycling

- explain the arrangements which the Waste Collection Authority expects to make with waste disposal contractors during the period of the plan

- identify the plant and equipment which it expects to provide for sorting and baling of waste

London Borough of Hounslow

The Audit Commission reported that Hounslow showed the fastest growth rate for recycling in London, reaching a level of 17% in 1998/99. This has been delivered through a combination of bring sites, a green waste trial scheme involving 4,000 households, and primarily through a multi-material kerbside recycling collection system.

The council has displayed considerable initiative in working with the community sector, and in making recycling services available to a broad range of their diverse community. This has included communication programmes for *hard to reach* groups, changing contractual arrangements for caretaker duties in high rise flats to facilitate recycling collections, and a neighbourhood intensive home composting programme with support and training.

The council is continually working with local partners, including a joint Real Nappy campaign with five other West London Boroughs, and promotion of the Buy Recycled campaign in conjunction with local supermarkets.

4.101 These plans should be reviewed and updated as necessary and should be taken fully into account by Waste Planning Authorities. Waste Collection Authorities should take account of relevant statutory performance standards and DETR/Welsh Office guidance issued in March 1998 to local authorities on preparing and revising recycling strategies and recycling plans. This guidance encourages close and effective liaison between the relevant Waste Planning Authorities, Waste Disposal Authorities and Waste Collection Authorities on waste management issues. Together with that guidance, the Secretary of State issued a direction that all Waste Collection Authorities in England should investigate if their recycling plan needed updating. As a result of this decision, the majority of Waste Collection Authorities in England are in the process of revising their recycling plans. In Wales, local authorities have been strongly encouraged to update their plans, in particular to take account of local government reorganisation in 1996.

A summary of waste collection routes

The rubbish bin is not the only outlet for domestic waste. While the availability of the following alternatives to the rubbish bin will vary across the country, the Government and the National Assembly are keen to see such initiatives being taken up by local communities more widely.

Almost all local councils have put in place *bring system recycling* banks which include bottle banks and can banks – some bring systems go further and can accept waste textiles, paper and plastic bottles. There are a number of advantages with bring systems:

- they can encourage high participation rates if sufficient banks, bins, etc can be located close to people's homes
- they are cheaper than separate collection from households

However there can also be some disadvantages, for instance badly sited banks can lead to an increase in traffic movements and may deny some people the opportunity to participate, and care has to be taken to ensure people put the right material in the right bin as contamination can damage the value of the collected materials.

An alternative to the bring system is *kerbside collection* of separated materials direct from households. A number of councils have introduced different kerbside collection systems. The most successful systems rely on the council working in close partnership with community groups, local businesses and charities, and others to achieve a good understanding across the local population of how the system operates and to ensure maximum participation rates.

Local authorities also provide *civic amenity sites* where people can bring their household waste – which consists generally of bulkier items such as furniture, DIY waste, kitchen equipment and garden waste, as well as recyclable waste.

People do not need to rely on the council to find alternatives to the rubbish bin: *reselling* things no longer needed at jumble sales and car boot sales remains a popular option, and charity shops often work closely with their local community to source a supply of clothes, textiles, books, toys and other objects for resale.

The possibilities of utilising home – or small community – composting schemes to handle kitchen and garden wastes is examined in more detail in Chapter 5 section 5.46 of this part of the strategy.

Manufacturers and retailers can also play their part – *take back schemes* can often prove to be a profitable addition to a product's marketing strategy, as can a proactive approach towards *goods maintenance* and repair agreements. Other sales techniques include the old-for-new *part-exchange* scheme where discounts are offered on new goods if the customer gives the company the product being replaced at the same time.

4.102   Recycling Credits are payments made by Waste Disposal Authorities in England to their Waste Collection Authorities. Payments are equal to avoided disposal costs, and support the Waste Collection Authority's recycling activities. The Government recognises that the current scheme might in some cases work against the closer working relationships which this strategy advocates. The Government is therefore proposing to examine with local government whether the financial incentives for the promotion of recycling are adequate.

4.103   For businesses, the Duty of Care requires them to make arrangements for the collection and safe disposal of their wastes. Commonly, companies will contract with waste management companies to collect and dispose of their waste, though on some industrial estates companies have entered into joint agreements with their local authorities to provide these services.

WASTE RECOVERY, TREATMENT AND DISPOSAL ACTIVITIES

4.104   Most waste generated in England and Wales is currently disposed to landfill. As the various actions to reduce waste generation, and to recover more materials and energy from the waste we do generate are bought to bear, our reliance on landfill will diminish. Our goal is to recover value from at least 45% of municipal waste by 2010 and to reduce industrial waste going to landfill to 85% of 1998 landfill levels, by 2005. The various waste management options are considered in more detail in Chapter 5 of this part of the strategy.

4.105   Waste Disposal Authorities are responsible for the safe disposal of municipal wastes arising in a particular geographical area. The Environmental Protection Act 1990 required Waste Disposal Authorities to transfer their waste disposal facilities to either an arms length Local Authority Waste Disposal Company or directly into the private sector, and to carry out their waste disposal responsibilities exclusively by means of letting contracts. The majority of Waste Disposal Authorities have now completed the divestment of their waste disposal operations as required by the legislation.

4.106   The Local Government Act 1999 applies the duty of Best Value to Waste Disposal Authorities. The Government's initial view is that statutory performance standards for recycling should apply to Waste Disposal Authority areas. The Government has recognised that the divestment provisions of the Environmental Protection Act 1990 do not sit well with the Best Value framework, which leaves the details of service delivery for the local authority to determine based on local circumstances. The Government has therefore announced that it intends to repeal these provisions.

4.107   Following the recommendation of the Review Group on the Local Authority Role in Recycling, the Government is actively encouraging Waste Disposal and Waste Collection Authorities to prepare joint Municipal Waste Management Strategies, and proposes to make these mandatory. The purpose of these strategies is to develop a strategic framework for the management of municipal waste. They will therefore set out, particularly in relation to contract specification, policies and proposals for the various collection, treatment and disposal options (such as waste reduction, re-use, recycling, composting, other forms of recovery and landfill). Strategies should take into account sustainability and local circumstances, with the aim of maximising environmental benefits, whilst at the same time minimising overall economic costs. They will also explain how these different practices can be developed in an integrated way.

4.108   Local authorities, in preparing these Municipal Waste Management Strategies, should consult with others with a legitimate interest, including the waste management industry, other affected local authorities, the local community and local businesses, and other relevant waste-generators. (One of the advantages of this approach is the extent to which local authorities are entering into partnerships with the private and community sectors). To date, about half of counties in England have already begun to develop Municipal Waste Management Strategies with their district councils, and most intend to have them in place before the end of 2000.

4.109   In Wales, authorities have been encouraged to consider wider waste management issues when revising their recycling plans, consulting with other local authorities and relevant organisations in the process. A number of authorities have already decided to incorporate their recycling plan into a broader waste management strategy.

INTEGRATING WASTE COLLECTION AND DISPOSAL ACTIVITIES

4.110   The Government believes that the current division of responsibility between Waste Collection and Waste Disposal Authorities can work against an integrated approach to waste management. Ensuring the development of such an integrated approach, and increased cooperation between local authorities (and their constituent communities and local businesses) will be a key factor in delivering both the objectives of this strategy and the challenging targets in the EC Landfill Directive.

4.111   Unitary Authorities already combine waste collection and waste disposal responsibilities in one body, and in those two-tier areas where Municipal Waste Management Strategies are being negotiated or have been put in place there is already increasingly close cooperation between collection and disposal authorities. The Government believes there is a need to build on this base, and will promote policies and initiatives aimed at improving cooperation and collaboration between authorities.

CHAPTER 5

Waste management options

5.1   The previous chapter considered the roles and responsibilities of those involved in waste generation, regulation, management and disposal. This chapter looks at the various waste management methods and techniques for managing waste.

5.2   The concept of the waste hierarchy suggests that the most effective environmental solution may often be to reduce the generation of waste. However, where further reduction is not practicable, products and materials can sometimes be used again, either for the same or a different purpose. Failing that, value should be recovered from waste through recycling or composting, or through energy recovery. Only if none of the above offer an appropriate solution should waste be incinerated without energy recovery, or disposed to landfill.

5.3   The structure of this chapter follows the structure of this theoretical waste hierarchy closely, with waste reduction, re-use, recycling, composting, energy recovery from waste, new and emerging energy recovery technologies, landspreading of wastes, and landfill each being considered in turn.

Waste reduction

5.4   Both the Government and the National Assembly are firmly committed to a strong emphasis on waste reduction. Unlike other options identified in the waste hierarchy, waste reduction is not an option that can be selected when we have no further use for a product. Rather, waste reduction needs to be kept in mind whenever we are making decisions about how to use our resources.

5.5   Successful waste reduction can often be the result of a dynamic interaction between producers and consumers. Businesses increasingly need to consider waste reduction options from the design of a product, through its manufacturing process, to the way it is transported, packaged and sold. Consumers also have to make an active decision to purchase goods to produce minimum waste.

Recognising waste generation and sustainable waste management opportunities

There are many situations where organisations and people simply do not recognise that waste is being generated, or that there are better, more sustainable options available for managing their waste.

For *manufacturers*, the problem is often to acknowledge that they are in fact generating unnecessary waste. There are a whole host of euphemisms that manufacturers use to describe materials not required in the final product, for example "inspection loss", "shop loss", "material variance" or "rejects". A simple way for manufacturers to gauge the amount of waste they need to deal with is to calculate the weight of materials used to produce goods, and subtract the weight of the goods produced. A significant proportion of the resulting difference in weights will be the amount of waste a production process is generating – and which needs to be dealt with sustainably.

Retailers face a similar problem to manufacturers – recognising that their activities can generate unnecessary waste. In particular, retailers can have a good deal of influence over the way the products they sell can be packaged, transported, stored, marketed and sold, which in turn will have an impact on the amount of waste generated in shops and warehouses.

Service providers – including those who concentrate on the retail and delivery of financial, leisure or similar packages – produce waste, both through their office activities and through the sale and maintenance of the packages they provide. For example, technological advances have led to improved office machinery such as computers and photocopiers, which in turn has led to obsolete equipment entering the waste stream – waste that service providers are capable of reducing, or managing more sustainably, through better contracts for maintenance and carefully designed upgrade strategies that take into account the uses to which redundant equipment (or their components) can be put.

Local authorities are very aware that waste is generated: they have a duty to collect the waste generated by householders and others, and then arrange for the management and disposal of that waste. A problem that many local authorities face is encouraging the community to recognise that waste management is not simply "the council's problem". While a local authority is legally obliged to manage its community's waste safely and in accordance with the principles of Best Value, they perform this service on behalf of the community.

Consumers – householders, schools, hospitals and others produce waste. The amount of waste consumers generate will depend entirely on the type of goods and services they purchase or use. All consumers have the power to influence the amount of waste they generate, by being aware of the amount of waste within the products they buy and use. Consumers can also make decisions to purchase goods and services which generate less waste, or more recoverable waste (for example, by buying wine in clear glass bottles rather than green bottles which are less easily recycled).

Similarly, by taking more care about how they dispose of their waste – through separating their waste into its constituent materials before collection – and by considering alternatives to the rubbish bin (such as municipal recycling facilities, product bring-back schemes, other local outlets for re-usable "waste" such as charity shops, home composting options and so forth) consumers can actively facilitate the implementation of more sustainable waste management systems (by local authorities and others) for the waste they generate.

ECODESIGN

5.6 Ecodesign is the process of producing more goods with less resource and less pollution, redesigning and re-manufacturing goods and services to enable recycling, and reducing harmful effects when they are returned to the environment. Producing more with less entails innovating in the way raw materials are extracted from the physical environment and used in the production process. This concept has been used by many companies to develop new working methods resulting in benefits both for their business and the environment.

INCENTIVES TO REDUCE HOUSEHOLD WASTE

5.7 The Government commissioned consultants to investigate the range of incentive schemes available that might be effective in encouraging householders to change their behaviour, such as reducing the amount of waste they put out for collection. The Government has decided to take forward further work on four types of schemes. These are:

- performance rewards – the householder receives some sort of financial payback or voucher proportional to the reduction in waste

- supermarket reward scheme – customers receive points on their loyalty cards or vouchers for products in exchange for recycling materials in the bring back schemes

- prizes for recycling – a competition is held where participating householders have the chance to win a prize

- intensive education – an intensive education programme to promote participation in waste reduction and recycling schemes

5.8 Work will be taken forward to pilot these schemes in a number of local authorities to determine how successful they can be in England. The National Assembly will consider how to take this forward in Wales.

Re-use

5.9 In the past, re-use played a prominent role both commercially and in the household: deposit-refund schemes and the doorstep delivery of products in refillable containers were widespread. Though the prevalence of traditional re-use systems has declined over the past few decades, a greater understanding of sustainable development across society is once again bringing the importance of re-use to the fore.

5.10 Some products are designed specifically to be used a number of times before becoming obsolete, for example, re-usable food and drink containers, rechargeable batteries and car tyres (which can be retreaded). With the introduction of the packaging Regulations, there is now an incentive for manufacturers to consider the use of refillable packaging, which will help them to reduce their costs and achieve their obligations under the Regulations. Bring-back schemes encourage the use of refillable containers, for example, for drinks such as milk and beer.

5.11 In other cases, goods can be refurbished or reconditioned to enable them to be put to the same use for longer or put to a different use. This might involve one or more individual components of an electrical product, for example. Reconditioned computer chips, although slower than the newer ones on the market, can be used in less demanding computer-controlled products. Or it could involve the whole product being refurbished to lengthen its life, for example refrigerators, cookers or furniture.

Regional Electronics Initiative

A number of community-based IT refurbishment centres across the Yorkshire and Humber region have been established and linked by the Regional Electronics Initiative, launched by Save Waste and Prosper Ltd. There are now computer refurbishment centres in Hull, Halifax, Bradford, Leeds and Scunthorpe, with others planned for Rotherham and Bridlington.

The centres have a number of common objectives:

- to divert electronic equipment away from landfill and towards re-use and recycling

- to provide low cost, quality guaranteed computer equipment to local and regional organisations that would not otherwise be able to obtain them, such as charities, voluntary groups and educational establishments, and to low income families

- to provide accredited training and full-time employment for local people

The refurbishment process varies depending on the type of equipment that is received. The equipment is logged and tested, and any data or markings which link the equipment to the donor are removed. When necessary, the machines are upgraded, then cleaned, tested and sorted for dispatch. New licensed software is installed to customer requirements. Any equipment which cannot be re-used is sent for recycling.

The Regional Initiative enables the centres to share good practice. It is working towards audited quality and environmental standards to provide further assurances for both donor and recipient. It also coordinates regional marketing of the centres, to promote and link them with other service providers, funders and businesses.

5.12   There is also the opportunity for everyone to make innovative use of used goods that would otherwise be discarded. Using plastic bags as bin liners, old clothes as cleaning cloths, and glass jars for storage are well known examples. Unwanted clothes and other items can be taken to charity shops. DIY television programmes provide ideas in decorating old items of furniture to give them a new lease of life.

5.13   Recent research[1] suggests that four out of five people re-use products. Plastic bags and glass jars or bottles are re-used by around half the public and plastic containers or bottles by one in five. There appears to be a strong tendency for those who recycle to also re-use items when they can.

5.14   There are good reasons for encouraging re-use, including:

- energy and raw material savings: many single-trip products are replaced by one re-useable one, thus reducing the need for the manufacture of new products

- reduced disposal costs

- costs savings for business and the consumer

- new market opportunities, for example in refillable products

[1] Waste Watch and NOP Research Group Ltd: *What people think about waste 1999*. ISBN 1 898026 72 6

SOFA project

There are 340 furniture re-use schemes throughout the UK which collectively receive over one million items per year for distribution to needy families. SOFA is larger than average but it represents the typical aims, objectives and activities of most other schemes.

SOFA is a registered charity and voluntary organisation, based in Leicester, which recycles and refurbishes items of furniture. It receives grant funding from a variety of sources including the local council. People wishing to donate items contact the project and a collection is made by the SOFA van. On average, 700 items are collected each month, covering a range of household items including kitchen appliances. These items are re-distributed to low income families by a referral procedure operated in association with the local social services and housing departments. A charge of less than £10 is made on each item.

5.15 However, there needs to be careful evaluation of the environmental benefits of enabling a product to be used more than once, set against the extra resources required to make it re-usable. For example, more energy and raw materials may be used in the production of refillable glass bottles, which will need to be more robust than non-refillable ones. There will also be costs associated with the need for infrastructure, including transport, for return systems, and with the washing and filling process. Other indirect impacts of re-usable goods can include the continued operation of less energy efficient products (such as refurbished refrigerators), and the lost market for the single use products replaced by re-usable products.

5.16 In particular, the number of times a product designed for re-use is actually re-used is important. If a refillable bottle, for example, is thrown away before the end of its useful life, it is likely there will be a greater use of resources than if the bottle had been manufactured for a single use.

5.17 Re-use systems should therefore be accompanied wherever possible by efforts to promote the environmental and economic advantages, both to consumers and businesses, in order to discourage disposal after one use. A change in behaviour is essential here, and education on the contribution that re-use can make towards sustainable development will be an important element in achieving that.

5.18 Re-using bottles can be particularly beneficial when the bottles can be recovered and refilled locally. The more often the bottle is returned and re-used, the more economical the system becomes.

Bags for life schemes

A number of supermarket chains now operate a *bag for life* scheme. The idea behind the scheme is to reduce the number of plastic carrier bags used by stores by providing a stronger, more durable bag. This *bag for life* lasts longer and, once worn out, is returned to the store to be exchanged for a new one. The old bags are then recycled.

The schemes are promoted to customers through leaflets available in stores. To ensure that the bags are not thrown away like conventional ones, customers pay a small charge for their first *bag for life*. Some supermarkets also offer customers a penny back each time the bag is re-used.

It is estimated that *bags for life*, being more robust than conventional *carry to car* bags, can last up to 20 trips. Since they also have double the carrying capacity, it is estimated each *bag for life* used can save the equivalent of 40 conventional bags.

Recycling

5.19   The benefits of recycling include:

- reduced demand for raw materials, by extending their life and maximising the value extracted from them

- making energy savings in the production process

- reducing emissions to air and water in the production process

- reducing disposal impacts, through less waste going to landfill

- promoting public awareness of environmental issues and personal responsibility for the waste we create

5.20   The Government and the National Assembly are committed to a substantial increase in recycling rates. This can only be achieved if action is taken in parallel on all three sides of the waste management triangle, namely:

- collection
- reprocessing
- markets

COLLECTION

5.21   Collection of recyclable materials requires an adequate infrastructure. This can be provided by the public sector (local authorities often in partnership with private or community organisations) or through contracts between businesses and waste management companies.

5.22   Separation of recyclable materials is an important element of collection. Comprehensive publicity and information campaigns, addressed at householders and individuals within companies, are essential, firstly to encourage them to separate out recyclable material from the waste in their bins, and secondly to ensure a high quality of sorted waste with the minimum of contamination by incorrectly placed materials. The Government and the National Assembly look to the respective industries, who have the expertise and knowledge, to take this forward.

British Steel CanRoute

CanRoute was launched by the British Steel Packaging Recycling Unit in June 1999 to address the challenge of collecting steel cans from households. It enables the economic collection of smaller quantities of steel cans from a large number of locations and is therefore particularly relevant to new community collection schemes.

CanRoute provides an essential link between collection in the community and recycling in the steel plant. With 13 regional centres around the UK, the system is designed to give collectors a delivery point as close as possible to the collection/sorting facilities, thereby minimising the cost and environmental impact of transportation. The regional CanRoute centre receives the cans, stores them and later supplies them to the steel making plant for recycling.

CanRoute provides collectors with a secure market for their cans at a competitive market price. During its first six months of operation, CanRoute has collected more than 64 million steel cans, with around 40 organisations signed up to the system, including local authorities, community organisations and small, independent collectors.

5.23   The House of Commons Environment Transport and Regional Affairs Select Committee, in their report on sustainable waste management, recommended that it was particularly important for waste collection authorities to ensure materials used to contain waste for collection and transportation were fit for purpose and, wherever possible, recyclable or biodegradable. The Government agrees with this recommendation: where containers made from recycled materials are practical, cost effective and make good environmental sense, they should naturally be preferred by local authorities.

REPROCESSING

5.24   The reprocessing industry has developed in a way dictated by the market, both in terms of technology used and in terms of capacity and location; for example reprocessing capacity for some materials is focused in only one part of the country. A significant expansion in production capacity for some materials will be required as part of a move towards a more sustainable waste management system. Policy instruments such as producer responsibility can be effective in stimulating the necessary investment.

5.25   Improvements in reprocessing technology can lead to cost savings, the ability to handle a wider range and quality of material, and an increase in the quality of the end product. Reprocessors will need to be aware of standards and specifications demanded by manufacturers, which may change as new applications are developed for recycled materials.

Alcan aluminium can recycling plant

Alcan opened its dedicated aluminium can recycling plant, the first of its kind in Europe, in 1991. The plant, in Warrington, Cheshire, produces aluminium ingots from used beverage cans. Each ingot contains approximately 1.5 million used cans. The ingots produced are transported to another mill to be rolled into sheet and then supplied to can makers for manufacture into cans again – a true "closed-loop" recycling system.

The plant employs the latest technology to maximise yield and energy efficiency. A four-stage process of shredding, de-coating, melting and casting takes place. Steel and other contaminants are removed during the shredding and de-coating operations. Technology employed during the melting process achieves rapid melting rates and high yields. The regenerative burners and burner management system reduce the energy costs by 30%, compared to an equivalent cold burner system. Each cast is tested for chemical composition and metal cleanliness, to achieve a high quality finished ingot. Waste gases from the reprocessing operations are removed from the plant and treated in a purpose-built emission abatement facility.

With a capacity of over 60,000 tonnes a year, the plant has the capacity to reprocess all the aluminium cans collected in the UK for the foreseeable future.

5.26   Through the Waste and Resources Action Programme (see Chapter 3 of Part 1 of this strategy), the aim is to facilitate investment in reprocessing. Incentives will be available both for existing industry, where capacity may already have been reached, and for the promotion of new reprocessors. The Programme will also aid reprocessing through the provision of technical and engineering advice and guidance, including a hands-on consultancy service, workshops and seminars.

Centriforce Products

Centriforce Products manufacture recycled plastic products for a wide range of applications in agriculture, utilities, industry and recreation. The company has developed a number of innovative processes, including blending high density polyethylene (HDPE) scrap with two specific fillers to create a wood-appearance sheet used in the construction of fence panels. HDPE scrap is also recycled into a range of "composite" profiles, used in pallets, furniture or decking, by blending the plastic directly with a recycled wood additive.

MARKETS

5.27   The Government in England has considered measures to stimulate the demand for secondary materials, both through its Market Development Group and through a research project examining the merits of policy instruments to secure an increase in demand. Work has also been undertaken by the Institute of Welsh Affairs into the development of recycling markets in Wales and their report *Waste in Wales – A National Resource: Generating Prosperity through Recycling* was published in February 2000.

5.28   The Market Development Group's key recommendation was that new markets for recycled goods and materials should be identified and developed as a priority measure; this new approach should be taken forward alongside other measures to develop and expand further the existing markets for recycled materials. The Group made a number of recommendations on these measures, which included environmental considerations in public procurement, market stabilisation measures, awareness and education, improved standards and specifications, best practice guidance and promotion of ecodesign.

5.29   The Government has welcomed the Market Development Group report and will be taking forward many of its recommendations through the Waste and Resources Action Programme. In particular, one of the Programme's primary objectives will be to develop markets and end-uses for secondary materials. Through its programme of research and information management, best practice advice and guidance, technical support, funding, market facilitation and training and development, it will also take action on many of the other recommendations of the Group, for example on best practice, standards and specifications, consumer awareness, education and ecodesign. The National Assembly will consider the implications of the Institute of Welsh Affairs report when taking forward the waste strategy in Wales.

Regional market development programmes

A number of regional programmes to develop the markets for recycled materials have been initiated throughout the UK.

As part of this network, for example, a Clean Merseyside Centre has been established in Liverpool. It aims to increase the local demand for recycled materials by:

- working with manufacturers to include recycled materials in their manufacturing process

- working with entrepreneurs to develop new products made from recycled materials

- facilitating joint ventures with local businesses to enable them to use recycling technologies which have proved successful elsewhere

The Clean Merseyside Centre has secured significant European Regional Development Fund and landfill tax credit funding, and will serve public and private sector organisations in the five local authority areas of Merseyside.

The programme will develop pilot projects to implement new ways of using recycled materials in new and existing manufacturing operations. Partnerships will be created between waste generators, collectors, processors and manufacturers in the region. The Clean Merseyside Centre will provide technical assistance and the business opportunities developed will be publicised through Business Links on Merseyside and through industry organisations.

5.30   In addition, the Government has already taken action on a number of the Market Development Group recommendations:

- we will pilot arrangements for a scheme to require public procurement of certain recycled goods, initially paper products

- on education and awareness, a high profile joint Government/industry seminar was held in March 2000 to promote awareness of the opportunities for using recycled plastics and to encourage manufacturers to use recycled plastics

- the Government has agreed new voluntary targets for the recycled content of newsprint with the Newspaper Publishers Association

---

Plastics Seminar

To follow up the recommendations of the Market Development Group, Government and industry jointly hosted a major seminar to promote the use of recycled plastics in March 2000. The event attracted over 160 delegates from a range of sectors, and set out to raise the low level of awareness among manufacturers of the potential uses of recycled plastic.

Delegates took part in sector-specific workshops to assess the potential for using the material. They learned about the huge new market opportunities for recycled plastics.

A conference report can be obtained from the British Plastics Federation (telephone 020 7457 5000). Government and industry will work together to take forward The Market Development Group recommendations, including promoting the opportunities for using recycled materials.

5.31　As a forerunner to the Waste and Resources Action Programme, the Department of Trade and Industry has launched a £1.4 million recycling programme aimed at securing and maintaining an increase in demand for secondary materials and providing an opportunity for businesses which make or use recycled materials to strengthen their competitiveness. The programme will fund a number of projects in line with these objectives.

Composting

5.32　Composting in England and Wales is a growing industry, and is acquiring greater significance as a waste management option. One of the key issues facing the industry over the next few years will be complying with the Landfill Directive targets in order to provide an alternative to landfill for biodegradable municipal wastes.

5.33　In May 1999, the Composting Association conducted its second annual survey to assess the state of the composting industry in the UK. The results show the progress made by the industry but highlight the need to increase composting capacity to meet the Landfill Directive targets.

5.34　The Survey indicated that there were 89 facilities operating in 1998, most of which were in England, composting a total of just over 900,000 tonnes. 59 of these were centrally-run sites, composting a total of over 800,000 tonnes. The remainder comprised eleven community operations, nine on-farm sites, nine on-site facilities and one miscellaneous site.

CENTRALISED COMPOSTING

5.35　Centralised composting is carried out where organic wastes are brought in from elsewhere such as a civic amenity site. These sites are mostly commercial operations, managed by specialist companies.

5.36　A comparison of the Composting Association's survey results for 1997 and 1998 suggested that the quantities of organic wastes composted at centralised sites across the UK in 1998 increased significantly (by over 500,000 tonnes). The survey also indicated that approximately 600,000 tonnes of municipal waste was composted in 1998, of which about two thirds (approximately 400,000 tonnes) comprised household waste. The majority (92%) of municipal waste comprised green wastes collected from civic amenity sites or local authority parks and gardens, with only 7% of organic municipal wastes collected at the kerbside.

5.37　If the composting industry is to meet the challenge of the Landfill Directive for municipal biodegradable waste, current capacities for composting this waste will need to be expanded significantly over the next decade. This increase will need to be even larger if the material deemed suitable for composting includes 50% of the paper and cardboard fraction.

Midland Composting and Recycling

Midland Composting and Recycling, a subsidiary of Jack Moody Holdings plc, operates a green waste collection service in the local authorities of Walsall MBC, Wolverhampton MBC, Dudley MBC and Birmingham City Council. In 1998-99, 15,000 tonnes of green waste were collected in these areas from kerbside collection schemes and civic amenity sites. This is expected to have increased to 40,000 tonnes in 1999-2000. The material produced, mainly mulch material and soil conditioners, is used for regeneration of urban or rural areas, for instance in new housing developments, and in highways schemes.

5.38   Garden waste, either from local authority parks or from civic amenity sites, makes up over 90% of all municipal waste composted. It will be critical to expand the amount of compostible waste collected at the kerbside in order to meet the targets of the Landfill Directive.

5.39   The survey also indicated that there were three sites composting in excess of 50,000 tonnes per annum. Economies of scale should also see the development of a greater number of larger sites in the future. The number of small-scale sites composting less than 5,000 tonnes per annum also increased (from 22 in 1997 to 35 in 1998). It is likely that these will be exempt from waste management licensing and probably involve community groups, on-site and on-farm facilities. A further increase in these small-scale, decentralised sites is expected in the next few years.

5.40   Finally, the survey indicated that there were 29 proposed new sites, composting an anticipated total of over 300,000 tonnes per annum. Of these, 11 were new centralised sites with a total expected throughput of just under 250,000 tonnes per annum, of which over 70% comprised sewage sludges.

St Edmundsbury Borough Council

St Edmundsbury Borough Council's increase in recycling from 2% in 1992 to 24% in 1998 owes much to the separate sorting, and kerbside collection of organic waste, combined with continued support for home composting. In 1995, following extensive trials of different systems, the council introduced a two-bin kerbside scheme for 9,400 households: one bin for kitchen and garden waste and one for the remaining refuse. Since then, the scheme has been expanded to cover 22,000 household and collects in excess of 7,000 tonnes of green waste for composting in a year. It is a joint partnership with Suffolk Waste Disposal and County Mulch, who have produced several products from the material; soil improver, surface mulch and general use lawn dressing.

BUSINESS AND COMPOST

5.41   A key objective for the Government and the National Assembly is to increase the amount of organic material in the waste stream which is composted. The retail sector has a major role to play. Some of the large supermarket chains are already running composting trials.

Supermarket composting

In 1997 Sainsbury's started a trial with the Organic Resource Agency (ORA) in Berkshire with funding from the landfill tax credit scheme. It involved traditional open air, regularly turned, compost heaps (windrows) using produce from their hypermarket in Calcot.

ORA monitored the quality and quantity of incoming material and the performance of the composting system, with fruit and vegetables being mixed with different combinations of cardboard and straw. The resulting compost was analysed for its nutrient content and tested in growing trials. It was found to make a good soil enhancer. However the most that might be composted was around one and a half tonnes of produce waste a week which means that more stores would need to be involved for it to be economically or commercially viable.

Sainsbury's is now working with ORA, Biffaward, Waitrose and an independent farmer to expand the trial across the central southern region. This will examine the costs and economics of these schemes more fully, monitor the performance of a range of different composting systems and establish the agricultural benefit of using compost. Sainsbury's have also linked with Alpheco Ltd. to have food waste from two Ipswich stores composted in large, temperature controlled, metal containers (an in-vessel system). Such schemes will also seek to demonstrate effectively that composting the putrescible waste from supermarkets and similar commercial sites is cheaper than disposal in landfill and can be operated in a way which has the least impact on the environment.

5.42   One initiative is a joint project by leading supermarkets known as *GROWS* looking at ways of maximising the opportunities to compost supermarket waste. The potential barriers include the health and safety implications of sorting waste in stores, the suitability of certain food products for composting (for example some foods may kill off the microbes responsible for decomposition), pathogen reduction and survival rates and the nutrient and mineral content of the product which may affect its end use.

5.43   In order to minimise the costs of transporting organic waste, composting may best be done at a local level. On the other hand composting sites need to be large enough to produce sufficient quantities on an economic basis. Operating a commercially successful composting operation will depend on getting the right balance between these two factors.

5.44   One way forward which would also keep the costs of land use and equipment to a minimum, would be for local businesses to operate in partnership with local authorities and the local farming community. This would create sufficient economies of scale at the local level and keep transport costs to a minimum. It may be for local authorities in some instances to take the lead if they have sufficient resources and expertise, although there is no reason why businesses should not do so too. When farmers are seeking to diversify their operations, there may be opportunities for them to host these schemes, particularly if they have sufficient land and the right equipment.

5.45   Compost for use in growing food crops is a form of closed loop recycling, and, if the compost is made to an appropriate standard, its use in agriculture will provide a reliable market for bulk quantities of processed organic wastes. It will also have a wide geographic spread which would encourage the establishment of local composting units. Comparative tests alongside proprietary (peat based) products have shown the beneficial use of waste-derived compost in promoting the growth of a number of food crops.

COMMUNITY COMPOSTING

5.46   Community composting is carried out by a group or groups of people in a local community, who pool their organic materials to make larger volumes of compost than is possible at home. In some instances, householders with small gardens or those who prefer not to manage their own compost use a local community project. An organisation called Community Recycling in Southwark Project (CRISP) has been piloting a range of methods of community composting on inner-city estates. In one particular example in a block of 44 householders, the kitchen waste is collected on a weekly basis. A local resident then mixes this into two large compost bins for use in the surrounding green amenity areas. Similar schemes have been run in various public institutions such as universities, hospitals, prisons and schools to manage the large amount of kitchen waste which they generate.

Devon Community Composting Network

This Group started off in 1993 with the launch of three projects, partly in response to the diminishing landfill space in the county, but also to help promote Agenda 21 principles generally. The Devon Community Composting Initiative evolved and developed into the Devon Community Composting Network. A small fleet of mobile chippers (including a shredder) visit individual local projects when enough material has been stockpiled, leaving the group, usually volunteers, to concentrate on making compost. The end result is that groups do not have to purchase, insure, store and maintain potentially dangerous machinery. Start-up grants are match-funded from other bodies; this usually pays for setting up the site, producing publicity material and so on. The Devon Recycling Committee also lends out shredders, trailers and rotary sieves. Some 1,700 tonnes of green waste will have been composted by March 2002, at an average cost of £15/tonne. This compares with hiring a skip for landfill which can cost up to £75 a tonne.

5.47   The Community Composting Network[2] (CCN) is a network of community composters which has over 125 members across the UK. They include community composting projects, local authorities and other supporting organisations. The CCN, a voluntary organisation, provides advice and support to new and existing community composting projects, promotes community composting at a national level, produces a quarterly newsletter and offers a range of information and advice.

HOME COMPOSTING

5.48   Home composting is a good opportunity for the householder to take responsibility for the organic fraction of their waste and provides an effective way of diverting it from landfill. Kitchen and garden waste (also known as putrescible waste) makes up about 21% of household waste, and there is considerable potential to compost this proportion of the average household's dustbin. However, it is important to emphasise that continued participation by householders over a period of time is the key to the success of home composting.

5.49   Home composting is very often promoted through the activity of the local authority. Some authorities provide subsidised compost bins, some also run *compost fairs* to raise the profile of composting at home. A survey of 324 UK local authorities in June 1999 showed that, since September 1995, 72% of them had made home compost bins available to their residents. Of those that had not, the three most common reasons given were lack of finance, greater focus on centralised composting and that responsibility had been passed to another branch of local government.

5.50   Authorities were also asked to indicate the number of compost bins they had distributed since 1995 and to estimate how many would be distributed during 2000/2001. By the end of 1999, there will be approximately 1.2 million home composting bins supplied by local authorities. This means that 5% of UK households (7% of households with a garden) will have a local authority compost bin.

5.51   In England, the majority of schemes, both existing and planned, are funded from local authority budgets. In Wales the majority of schemes are funded from local authority discretionary budgets.

---

2 Community Composting Network, 67 Alexandra Road, Sheffield S2 3EE

Successful home composting

To successfully compost, the right mix of organic matter, soil, water, warmth and air is needed. Most types of organic material can be added to the compost heap, although meat and dairy products, fat, bones, sugary foods, weeds, diseased plants, or animal manure should be avoided. For best results, approximately equal amounts of greens and browns is recommended – for example fresh organic materials such as kitchen scraps, green leaves, tea bags, with dead organic matter such as straw, sawdust and dry autumn leaves. A small amount of soil should be mixed into the heap to introduce the bacteria it needs to turn it into compost. Sufficient water should also be added as necessary to ensure the material is kept moist. The compost heap should be covered during cold weather to insulate the bacteria and it is advisable to stir the heap regularly to speed up the composting process and to stop it becoming too smelly.

More detailed advice and information about home composting is available from HDRA – the organic organisation, at Ryton Organic Gardens, Coventry CV8 3LG.

WORMERIES (VERMICULTURE)

5.52   A wormery is container for kitchen and garden waste which houses a colony of worms that break down the organic material to produce very fertile compost. Wormeries can be kept indoors or out, and are ideal for households with no gardens as they thrive on kitchen scraps, shredded newspaper and cardboard. You can make your own wormery in a bin, the size of a regular bucket; you need special compost worms known as brandlings, which can be obtained from fishing shops. Like all livestock they need to be cared for and kept happy. They need a moist but not waterlogged environment and need to be fed little and often.

Green Business Network

Green Business Network have developed a scheme using wormeries to compost waste cardboard. The Network includes Kirklees MBC, Calderdale MBC and Business Link. Trials with local stable owners had found shredded cardboard to be a very successful alternative to conventional animal bedding materials such as wood shavings or straw. Shredders at Pennine Magpie and Vocational Enterprises (a recycling charity providing vocational training for people with learning difficulties) shred the waste cardboard, which is then transported to Huddersfield Community Farm and local stables. Once mixed with animal manure (an ideal activator), the used bedding is composted by vermiculture, and broken down into high quality odourless and peat free compost. Following the success of the first trials, a further scheme was established at the Ponderosa Rural Therapeutic Centre.

Waste as a fuel

5.53   There are four broad ways in which energy can be recovered from waste:

- direct waste incineration

- use as a fuel substitute – either directly or as refuse derived fuel

- materials recovery, with energy released as a by-product of the process (for example anaerobic digestion)

- waste disposal, with fuel recovered as a by-product of the process (such as landfill gas)

5.54   Where energy recovery forms part of an integrated waste strategy, the potential for incorporating combined heat and power (CHP) technology should always be considered in order to maximise the energy which is recovered.

5.55   Energy from waste can make an important contribution towards sustainable development as a source of renewable energy, reducing the use of fossil fuels and cutting emissions of greenhouse gasses. The Government's commitment to renewable energy is reflected in its recently published paper – *New and Renewable Energy – Prospects for the 21st Century*, which sets out the Government's position on renewable energy, including energy from waste. In Wales, the National Assembly is producing a strategic framework for energy developments.

5.56   Furthermore, the Government underlined its commitment towards using waste as a source of renewable energy when it announced the results of the bidding for the Fifth Non Fossil Fuel Obligation Order, accepting bids for electricity generation to be powered using waste, either directly or in the form of refuse derived fuel. Other successful bids included proposals to generate waste from burning sewage sludges, agricultural wastes and methane generated in landfill sites.

New and renewable energy

The Government undertook a Manifesto commitment to a new and strong drive to develop renewable sources of energy. A consultation document, *New and Renewable Energy – Prospects for the 21st Century*, was published in March 1999. The Government published its conclusions in response to the public consultation in February 2000.

The response introduced the four key strands of the new and renewable energy strategy:

- powers under the Utilities Bill for an Obligation on all electricity suppliers to provide an increasing proportion of their power from renewable sources

- exemption of renewable-generated electricity and heat from the Climate Change levy

- an expanded new and renewable energy support programme, including research, development, demonstration and dissemination of information

- development of a regional strategic approach and targets for renewable energy

The Government believes that it will be essential to ensure that commercially viable projects already contracted under the Non-Fossil Fuel Obligation (NFFO) arrangements continue to attract and retain project finance. Accordingly, transitional arrangements which aim to ensure that these projects – which include a number of energy from waste projects – continue to be viable are being developed by the contracting parties. The Utilities Bill will give powers for secondary legislation to be developed to facilitate this process.

INCINERATION WITH ENERGY RECOVERY

5.57   If we are to achieve a sustainable waste management system, then incineration with energy recovery will need to play a full and integrated part in local and regional solutions developed over the next few years. Waste to energy incineration must be considered in the context of an integrated approach to waste management which encourages waste reduction, re-use and recycling. Where incineration with energy recovery is the best practicable environmental option, the potential for incorporating combined heat and power should always be considered in order to increase the efficiency of the process. Energy from waste schemes will be given a boost by the exemption of renewable energy and of good quality combined heat and power from the Climate Change levy.

Combined Heat and Power

Combined heat and power (CHP) is a highly fuel efficient technology which produces electricity and heat from a single plant. When electricity is generated only a part of the input energy is converted into electricity (typically 30-50%). The remainder of the energy consumed by the generation is dissipated via cooling systems as waste heat. If a suitable use for this heat can be found, for example in nearby buildings, the heat can be recovered, raising the efficiency of the process to as much as 80-90% and saving up to 40% on fuel bills.

Whilst the majority of existing CHP plants use conventional fuels, such as gas, coal and oil, a growing number are using bio-fuels and by-products to fuel the process. Such alternative fuels include municipal, industrial or clinical waste; landfill gas; poultry litter; sewage sludge; and other bio-fuels such as energy crops.

The Government and the National Assembly are working with industry to promote the wider uptake of CHP because of its environmental and economic benefits. Every 1000 megawatts of CHP can reduce energy costs by £100 million and carbon emissions by around 1 million tonnes per annum. CHP will be a key element in achieving our aims under the legally binding Kyoto Protocol which requires the UK to reduce its greenhouse gas emissions by 12.5% by 2008-2010, and to move towards our domestic goal of reducing carbon dioxide emissions by 20% by 2010.

Studies by the Government's Energy Technology Support Unit (ETSU) have indicated that there is an economic potential for between 10 and 19 thousand megawatts of CHP in the UK in industry, commercial, residential and community heating sectors. We are actively working towards a new target of at least 10,000 megawatts of installed CHP capacity by 2010.

Combined heat and power schemes incorporating municipal solid waste incineration enable the energy efficiency of the process to be increased to as much as 75%, compared with the normal efficiency of around 25-35% from a conventional energy from waste plant.

Municipal waste incineration with energy recovery – combined with CHP plant and a community heating system – has the potential to provide an integrated, sustainable and cost-effective means of managing waste locally, particularly in certain urban locations with a large heat demand nearby, and supplying affordable power and warmth for the community.

There are waste to energy schemes incorporating CHP and community heating schemes at Sheffield, Nottingham and Newcastle. A further 17 energy from waste and CHP projects have received awards under the Non-Fossil Fuel Obligation (NFFO) rounds 4 and 5 in 1998 and 1999; and a number of existing energy from waste schemes have the potential to be enhanced to achieve full CHP and community heating potential, for example SELCHP in South East London.

5.58 By far the most common method of energy recovery is by incineration in specially designed facilities. Incineration facilities can include *mass burn systems* which burn municipal solid waste with little pre-treatment, and are usually large (taking more than 500 tonnes of waste each day, or 200,000-400,000 tonnes a year), to gain economies of scale. *Modular burn systems* are smaller (usually taking between 50 and 250 tonnes per day of waste, which is between 20,000 and 90,000 tonnes per year) and are designed for use by local communities.

5.59 Waste incineration is one of the most technically highly developed waste management options available at this time. There are a wide variety of combustion systems developed from boiler plant technology and also more novel techniques such as *molten salt* and *fluidised bed incinerators*. It is also among the most strictly regulated waste management options, with a current proposal to replace existing European controls with a new Directive which will apply stringent emission limits to virtually all types of waste incinerator.

5.60 The fuel used in incinerators tends to be *municipal solid waste* (MSW), but could include other carbon-based waste streams with calorific value. It used to be the case that municipal solid waste was burned as delivered, but incinerators increasingly use other recovery processes either before incineration (materials recovery facilities) or after incineration (metals recovery, and use of bottom ash as a construction material).

5.61 Energy is recovered from incineration in the form of *heat*, which is used to generate high pressure *steam*, which in turn can generate *electricity*. Around 10% of electrical energy is used in-house while the remainder is exported to the countrywide electricity grid. With conventional power facilities the waste heat from the turbine is discarded. In *combined heat and power plants* this residual energy is recovered for use in for example, community heating schemes, thus improving the thermal efficiency of the process (from between 25-35% up to as much as 75%) and providing a further source of revenue.

Nottingham Combined Heat and Power and community heating scheme

An energy from waste scheme at Nottingham was started back in the 1960s to help deal with a shortage of landfill space for the city's municipal waste. The Council decided to put the waste to use in an incinerator to produce heat for local homes and businesses. In the 1990s, the scheme was refurbished and, with support from NFFO, upgraded to CHP to generate electricity.

Today the Nottingham scheme is a state of the art, Energy from Waste CHP/Community Heating scheme. The municipal waste used in the scheme reduces the volume of waste going to landfill by 70% and cuts carbon dioxide emissions by 34,000 tonnes a year. By using waste 300 Gigawatt hours of fossil fuel is saved each year.

The scheme now provides low cost heat to over 4,000 local authority and housing association homes and around 700 private homes. Mixed in with this are a number of commercial and public buildings including Trent University and the Victoria Centre (a 120-outlet shopping centre). The Nottingham scheme also delivers lower maintenance and capital costs for landlord and tenant, along with reliable and economic operation.

ENSURING SAFER WASTE INCINERATION

5.62 Pollution is an issue of vital importance. People have every right to demand the highest standards from all waste management facilities. The setting and enforcing of high environmental protection standards is a priority for the Government and the National Assembly.

5.63 The environmental performance of waste incinerators has improved immensely since the early 1990s. In 1992 there were 32 municipal waste-only incinerators. Following the agreement in 1989 of tight controls on incinerator emissions in Europe, and the imposition of UK-wide dioxin controls, most municipal waste incinerators in England and Wales were closed. A few facilities were upgraded to meet these tough standards, and new incinerators built to comply with the standards have come on-line over the past 2-3 years. The proposed Waste Incineration Directive includes even tighter controls on emission limits, as well as measures to ensure safe and efficient operation of incinerators.

FUEL SUBSTITUTES

5.64 Some industrial processes and power producing facilities operate under conditions which allow the use of high calorific value waste in place of conventional fuels. The most common example is cement manufacture where high temperatures and long residence times ensure complete combustion of the waste; the highly alkaline conditions in the kiln remove acid gases and metals from the gas stream; and the ash is retained in the clinker. In this case, the benefits to the environment go hand-in-hand with a helpful contribution, through fuel cost reduction, to improving the international competitiveness of the cement industry.

5.65   Typical wastes which are burned in these processes include municipal solid waste, tyres and spent solvents. Wastes can be burned untreated, but more usually solid wastes are shredded to assist feeding the material to the kiln. Solvent wastes are required to meet a specification to limit the quantities of potential pollutants such as PCBs, and to ensure that the waste is being burned primarily for its calorific value rather than as a disposal option.

5.66   Integrated Pollution Control (IPC) authorisations limit the extent to which processes can substitute fuels with waste. The draft Waste Incineration Directive also includes limits on emissions which will be placed on plants burning waste alongside other fuels.

Refuse derived fuel

5.67   Refuse derived fuel (known as RDF) is usually produced from municipal solid waste with recyclable and non combustible materials removed. The product comes in two varieties: either loose (coarse RDF), or compressed into pellets (densified RDF). Refuse derived fuel was originally developed as a substitute fuel for use in coal fired boilers, but it can cause fouling of the boiler tubes which reduces energy efficiency, and the markets for solid fuels is much reduced as facilities have been converted to gas or oil in preference to coal. There are currently three facilities producing densified RDF in the UK.

Production of energy through waste disposal

5.68   Landfill gas is a methane-rich biogas (typically 65% methane and 35% carbon dioxide) formed from the decomposition of organic material in landfill. The gas can be used to fuel reciprocating engines or turbines to generate electricity, or used directly in kilns and boilers. The quality of the gas and level of impurities varies depending on the mixture of waste in the landfill, the degree of decomposition and the age of the site. In some countries the gas has been cleaned to pipeline quality or used as a vehicle fuel.

5.69   Methane is a powerful greenhouse gas which landfill operators are obliged to collect and treat either by flaring or through energy recovery. Landfill gas energy recovery schemes have been supported by the Non Fossil Fuel Obligation. Since the introduction of NFFO in 1990, 306 projects have been contracted with a planned capacity of 650 megawatts of power, of which 200 megawatts (representing 107 projects) were operational at the end of 1998[3].

5.70   It has been proposed that there may be value to be gained in excavating landfill sites. One of the first instances of this practice was in the United States of America, to recover non-decomposed material with thermal value to provide additional fuel for a waste to energy plant. However, it is unlikely that recovery operations alone would provide sufficient justification for landfill mining in England and Wales since our landfills are designed and operated to encourage decomposition. American case studies have shown that through mining up to 60% of landfill space may be recovered for further use. Recoverable materials are mainly soil for use as top cover and material suitable for recycling as aggregate. Metals recovery is generally not feasable as the material requires substantial processing to improve its quality for sale.

5.71   However, it is possible that landfill mining could be an option where it was necessary to restore closed landfills for an alternative use, or to extend the life of an existing site, or as part of remedial works to upgrade a polluting site.

3 Digest of United Kingdom Energy Statistics 1999

New and emerging energy recovery technologies

5.72   If our waste management system is to become sustainable, then we need to adopt an integrated and multi-faceted approach towards waste management. Our immediate need to divert waste away from landfill mean that much of our effort over the next 5-10 years is likely to concentrate on increasing our recycling, composting and incineration capacity. But these are not the only waste management options that exist. Innovative technologies could be developed and (if they prove to be safe, viable and competitive compared to other waste management options) might take their place alongside the more traditional options during the lifetime of this strategy. A few of these options are highlighted below.

5.73   **Pyrolysis.** In this treatment, organic waste is heated in the absence of air to produce a mixture of gaseous and liquid fuels and a solid inert residue (mainly carbon). Pyrolysis generally requires a consistent waste stream such as tyres or plastics to produce a usable fuel product. Currently, there is only one facility established in the UK – taking in tyres.

5.74   **Fermentation.** This particular treatment is confined mainly to agricultural wastes, but can in theory also be extended to pre-treated municipal solid waste, to produce liquid fuel (ethanol, and some methanol).

5.75   **Anaerobic digestion** is the biological degradation of organics in the absence of oxygen, producing methane gas and residue (digestate) suitable for use as a soil improver. It has been used successfully for many years to treat sewage sludges; the methane gas is used to meet on site power and process heat requirements. It has also been used to treat cattle slurry on farms. It is possible that the process can treat the organic fraction of municipal solid waste, but there are reservations about the cost and the high degree of segregation required to produce a marketable digestate.

5.76   **Gasification** is where carbon based wastes are heated in the presence of air or steam to produce fuel-rich gases. The technology is based on the reforming process to produce town gas from coal, and requires industrial scale facilities. From the end of 1998 the dumping of sewage sludge at sea has been prohibited. In response to this Northumbrian Water has proposed gasification for the treatment of sewage sludge in its area.

5.77   **Feedstock recycling.** There are problems associated with the recycling of plastics. The high volume to weight ratio makes collection and transport to centralised points uneconomic, post-consumer plastic packaging is usually contaminated and – owing to the wide range of types of plastic in use – segregation of compatible materials is a problem. However it is possible to react mixed plastic waste in a polymer cracking process which results in a hydrocarbon product similar to the naphtha feedstock used in petrochemical plants for the manufacture of bulk plastics. Ideally the process should be sited adjacent to existing petrochemicals facilities with bulk plastics waste being shipped in by rail. Such a process could contribute effectively to the plastics recovery rate and would be true closed loop recycling.

5.78   **Feedstock substitution**. Another potential application is the use of mixed plastic waste as a feedstock in blast furnaces producing pig iron. Here coal, oil or natural gas (which are essentially a convenient source of carbon) act as a reagent to reduce iron ore to the metal. Mixed plastic waste can be used as a substitute source of carbon. The process has been adopted by the iron and steel industry in Germany, reportedly using 100,000 tonnes of waste plastic for this purpose in 1996.

5.79 **Substitute fuels.** Many wastes contain materials which have a high calorific value and can be suitable for use as substitute fuels in industrial processes. There are cases in common use already in England and Wales such as scrap tyres and solvent wastes which substitute coal and petcoke in cement and lime kilns. European experience has demonstrated that packaging waste paper, biofuels and plastics can all be used as substitute fuels in cement kilns.

5.80 **Plasma arc.** Municipal solid waste incineration typically reduces the volume of waste by 90%. However this volume reduction is compromised by the additional waste generated in cleaning the exhaust gases which are contaminated and require treatment. This normally involves the addition of lime to neutralise acids, and carbon to remove residual organic species such as dioxins followed by filtration to remove particulates (fly ash). Around 30% of the capital costs of a conventional incineration facility is attributable to the flue gas clean-up system. This is likely to increase significantly as tighter discharge limits require the installation of additional treatments. Gas treatment residues are treated as hazardous waste.

5.81 Alternative heat combustion systems for mixed wastes such as municipal solid waste, adapted from processes already operated in the metal refining industry, are being developed. These avoid the need for the large volumes of air required to support combustion. These typically use plasma arc heating (the energy released by an electrical discharge in a inert atmosphere) to raise the temperature of the waste to anything between 3,000 – 10,000°C, converting organic material to a hydrogen rich gas and non combustibles to an inert glassy residue. The gas (which is relatively uncontaminated) is suitable for generating electricity to support the process. The volume of gases discharged from these processes is generally less than 10% of that generated by incinerators with the same waste processing capacity.

Biomechanical waste treatment

5.82 Biomechanical waste treatment (BWT) is a generic name for a range of processes designed to recover valuable components from unsorted municipal solid waste for recycling, and deliver a stabilised residue for final landfilling.

5.83 Biomechanical waste treatment commonly comprises of a number of standard waste separation operations to remove recycled materials such as glass, metals and plastics, followed by composting of the remaining organic materials. Variants of the process use anaerobic digestion instead of composting. Biomechanical waste treatment does not aim to generate marketable compost since the required quality can only be achieved using source separated organic wastes. However, it is claimed that the potential of the residue for landfill gas production is reduced by up to 90% and the quantity of residual waste disposed to landfill is reduced by 40-60%. In this way the lifespan of landfills can be extended, and the process would contribute to meeting the targets of the Landfill Directive.

5.84 It is envisaged that biomechanical waste treatment could be part of an integrated waste management system that incorporates waste reduction activities and source separation schemes for recyclable materials, with the remaining mixed waste treated biomechanically.

5.85   The system is in common use in Austria and Germany, where limits on the disposal of organic waste to landfill have been in place for some years. However, it has been reported that the process can result in unacceptable levels of volatile organic compounds, ammonia, methane and heavy metals being emitted into the atmosphere. The Environment Agency would have to be satisfied that no excessive emissions were produced before licensing an individual facility.

Landspreading waste

5.86   Landspreading (or land treatment) represents an economical and – when properly controlled – environmentally safe way of recovering value from a variety of organic wastes. Most agricultural wastes and by-products are organic – for example, manure, slurry, silage effluent and crop residues – and landspreading is the normal waste management option for these materials. Sewage sludge and certain industrial wastes – for example, paper sludge, food processing waste and non-food waste such as lime and gypsum – may be spread on land beneficially. The Framework Directive on Waste classifies landspreading as a waste recovery operation – *land treatment resulting in benefit to agriculture or ecological improvement.*

5.87   All the above wastes provide valuable nutrients which allow farmers to reduce the amount of inorganic fertiliser applied, and can lead to improvements in soil structure. The Government proposes to consult on a draft soil strategy for England in the near future, while in Wales the National Assembly is considering how a similar strategy should be developed. It is a response to the leading recommendation in the Nineteenth Report of the Royal Commission on Environmental Pollution, *Sustainable Uses of Soil*, which describes how soil is a resource that is often taken for granted. The strategy will bring together the many individual policies and activities which contribute to soil protection and improvement within a single coherent and comprehensive strategy. It will be one of the follow up documents to the Sustainable Development Strategy, and its overall aim is to promote the sustainable use of soil.

5.88   There are also potential disadvantages to landspreading wastes. Used inappropriately, landspreading may lead to soil contamination from the concentration of some elements, may lead to deterioration in soil structure, may produce offensive odours and may pollute water (including groundwater).

5.89   Under the Waste Management Licensing Regulations 1994, the landspreading of certain wastes is exempted, as a waste recovery operation, from waste management licensing controls if it complies with certain rules. The exemption applies if the landspreading of the waste will result in benefit to agriculture or ecological improvement, no more than a specified amount of the particular waste is applied to each hectare of land, and the Environment Agency is informed in advance of the proposed landspreading. The requirement to demonstrate ecological improvement guards against the potential disadvantages described above.

5.90   The Government, together with the Environment Agency, commissioned research to develop further the criteria that determine whether the landspreading of waste benefits agriculture or results in ecological improvement. In the light of findings of this research, the Environment Agency made recommendations in 1999 on the future regulation of landspreading within the Waste Management Licensing Regulations. These recommendations are being considered and any proposals to revise the current licensing exemption will be published for consultation. The Regulations and guidance (as at 2000) are as follows:

- the Waste Management Licensing Regulations 1994 (as amended)

- the landspreading of organic farm waste is covered by *Codes of Good Agricultural Practice for the Protection of Water, Air and Soil*

- the landspreading of sewage sludge on agricultural land is controlled by the Sewage Sludge (Use in Agriculture) Regulations 1989 (as amended). These Regulations implement EC Directive 86/278/EC and are complemented by a Code of Practice

Landfill and landraising

5.91   The new European Directive on the landfill of waste will have a significant impact on landfill practices in the UK. We have until July 2001 to transpose the Directive into UK law. It will ensure that landfill sites across the European Union face strict regulatory controls on their operation, environmental monitoring and long-term care after closure.

5.92   The Government and the National Assembly welcome the Directive. Indeed many of its controls are similar to those already in existence in the UK under our Waste Management Licensing Regulations 1994 and Environmental Protection Act 1990. The Directive will also bring in some key changes. For instance, the Directive will require a landfill site to be categorised into one of three types depending on the type of waste it receives: hazardous; non-hazardous; or inert. This effectively ends the traditional UK practice of co-disposing hazardous waste with non-hazardous waste, and will require the modification of landfill sites and the expansion of alternative means of dealing with hazardous wastes. The Directive will also require the treatment of wastes before landfilling, for instance to reduce the hazardousness or volume of the waste being landfilled. It will also ban the landfill of certain types of waste, for instance liquid wastes, tyres, and certain hazardous wastes. The changes needed to implement these controls will be the subject of a consultation paper due out in 2000. This follows seminars held by DETR and the Environment Agency in February 2000, which provided a forum for discussion of these issues.

5.93   Another main objective of the Landfill Directive is to reduce the emission of methane, a powerful greenhouse gas, from landfill sites. The Government and the National Assembly take very seriously their obligations to tackle climate change. This is the most serious environmental problem facing the world and the UK is continuing to lead efforts to tackle it. Currently, more methane gas collected at landfills is burnt off rather than used for energy, and the Environment Agency is actively addressing this issue. These steps are all the more essential since the Directive will require the collection, treatment and use of gas from all landfills receiving biodegradable waste.

5.94   The new Directive also introduces progressively diminishing limits on the landfill of biodegradable municipal waste. Introducing these reductions and implementing the changes that will be required to meet them presents a significant challenge to us all: Government; industry; and the wider public. This will require the full implementation of the measures to expand the alternative means of dealing with biodegradable waste set out in the sections dealing with recycling, composting and energy from waste, and the measures to reduce the amount of waste we produce set out in the section dealing with waste reduction. The formal mechanism for implementing the biodegradable municipal waste targets in the Landfill Directive is set out in paragraphs 5.100 onwards.

5.95 Reducing the amount of waste going to landfill is compatible with its continued use as a viable waste management option. Properly regulated landfill can go some way to mitigating the environmental impact of disposing of waste to land, for example methane generated from the decomposition of waste in landfill sites can be collected and used to generate power, and landfill can be a means of restoring full landscape use after activities such as quarrying. Moreover, landfill will remain the Best Practicable Environmental Option for certain wastes in certain circumstances, and indeed may be the only option for wastes such as heavy sludges from some industrial processes.

5.96 The position of the Government and the National Assembly on landraising – disposing of waste on top rather than below ground – is similar to that on landfill. Although a much less common practice this has the same environmental disbenefits as landfill and generally does not share the benefit of land restoration.

5.97 The Government is committed to reducing our reliance on landfill through the Landfill Tax. This tax on the disposal of waste to landfill was introduced in October 1996 at a rate of £7 per tonne for active wastes and a lower rate of £2 per tonne for inactive wastes. The tax was designed to promote the *polluter pays* principle by increasing the cost of landfill to reflect its environmental costs, and to promote a more sustainable approach to waste management in which less waste is produced and more is recovered or recycled. From April 1999, the Government increased the rate for active wastes to £10 per tonne, and is committed to continuing to increase it by £1 per tonne per year, with a review in 2004.

5.98 The Landfill Tax Credit Scheme allows up to 20% of the funds generated by the tax to be channelled into bodies with environmental objectives. The aims of the scheme are:

- to help promote and foster sustainable waste management practices which provide an alternative to landfill

- to help projects which benefit communities in the vicinity of landfill sites, thereby helping to compensate for the disamenity effects and environmental impact of landfill

Proposals for using the scheme to fuerther help recycling are set out in Chapter 2 of Part 1.

CONSULTATION PAPER *LIMITING LANDFILL*

5.99 In October 1999 the Government and the National Assembly for Wales started the process of transposing the EC Landfill Directive into UK law, with the publication of *Limiting Landfill* – a consultation paper on limiting landfill to meet the Directive's targets for the landfill of biodegradable municipal waste.

Instrument to limit landfill

5.100 The EC Landfill Directive sets specific targets for the reduction of biodegradable municipal waste sent to landfill over the next two decades. *Limiting Landfill* consulted on five options for an instrument to limit landfill. These options were:

- Option 1: a ban on the landfill of biodegradable municipal waste

- Option 2: a ban on the landfill of certain types of biodegradable municipal waste

- Option 3: permits for landfill operators to accept biodegradable municipal waste

- Option 4: permits for local authorities to landfill biodegradable municipal waste

- Option 5: the Landfill Tax

5.101   In all 203 responses were received. Of these, the overwhelming majority – over 70% – were in favour of Option 4: permits for local authorities to landfill biodegradable municipal waste. When responses were analysed by sector (local authorities, the waste industry, householders, associations, and commercial organisations), the preferred option within each sector was Option 4. Overall there was some support for Option 3 (15%), but support for Options 1, 2 and 5 was less than 5% in each case. The main reason cited by respondents for supporting Option 4 was that this option allowed local authorities to retain control over planning for the disposal of municipal waste.

Further consultation on instrument to limit landfill

5.102   The DETR has held wide-ranging discussions with industry, local authority sounding groups and other government departments. These have also confirmed support for Option 4.

5.103   Further work by independent consultants provided support for Option 4. Their conclusions were that under Option 4:

- total compliance costs are likely to be least where waste disposal authorities hold permits to landfill biodegradable municipal waste, and these are tradable

- there is least transfer of wealth into or outside the local authority system (providing that permits are granted free and not auctioned to local authorities)

- Administrative costs are likely to be lower than other options, as fewer and more homogeneous participants are involved in this option than would be the case with other options

5.104   As outlined in Part 1 of this strategy, we therefore propose to introduce, as soon as the legislative programme allows, a permit system to limit the level of biodegradable municipal waste local authorities can send to landfill over the next two decades.

Tradability of permits

5.105   In general, respondents to Limiting Landfill felt that permits should be allocated to authorities by central government, rather than auctioned, and that authorities should be free to trade permits to achieve compliance at least cost.

5.106   We therefore propose to introduce a system of permits for local authorities to landfill biodegradable municipal waste, that can be traded between parties.

5.107   We will conduct further consultation on the design of this permit system as part of consultation on the draft implementing legislation.

Sanctions for failure to comply with provisions of permit

5.108　Respondents typically felt that sanctions needed to be significant to be effective, and that any fines for breach of permits should be set at a level higher than the cost of achieving compliance. Fines, forfeiture of functions, and naming and shaming were amongst suggested sanctions.

Incineration without energy recovery

5.109　Incineration without recovery of power or heat is categorised as waste disposal and is not an option that the Government and the National Assembly would generally wish to encourage for non-hazardous wastes. Incineration of hazardous wastes is discussed in Chapter 6 section 6.23 of this part of the strategy.

CHAPTER 6
Handling hazardous waste

6.1   The Government and the National Assembly will work with industry to reduce both the amount of hazardous waste generated and the hazardousness of waste. Hazardous wastes pose particular risks to health and the environment, so it is especially important that they are managed properly.

Dealing with hazardous wastes

6.2   The need to adopt a more strategic approach to hazardous waste management has been reinforced by the need to implement the Landfill Directive and the IPPC Directive as well as the recently published Chemicals Strategy and specific measures to remove the most dangerous chemicals from the environment. Furthermore, in December 1999, the Fifth Conference of Parties to the Basel Convention made a high-level declaration on the environmentally sound management of hazardous waste. As a Party to the Convention, the UK will attempt to follow the principles of this declaration which encourages both capacity building in developing countries and hazardous waste reduction in developed countries.

What is hazardous waste?

6.3   The Hazardous Waste Directive (91/689/EEC) provides the framework for the control of hazardous, or "special" waste. The annexes to the Directive include descriptions of categories or generic types of hazardous waste according to:

- their nature or the activity by which they were generated and

- properties of wastes that render them hazardous

6.4   Furthermore, a European Commission Decision (94/904/EC) established the Hazardous Waste List. Wastes are listed because they display one or more hazardous characteristics or properties such as being explosive, highly flammable, toxic or carcinogenic.

6.5   Household wastes that have hazardous properties, such as bleach and some batteries, are at present excluded from the Directive. However, the European Commission has signalled its intention to propose a new Directive dealing with household hazardous wastes.

6.6   The Special Waste Regulations 1996 (SI 1996 No. 972, as amended by SI 1996 No. 2019) implement the Hazardous Waste Directive. They ensure that special wastes are tracked, through a consignment note system, from the point at which they arise until they reach the facility where they are to be recovered or disposed of.

6.7 The 1996 Regulations replaced the Control of Pollution (Special Waste) Regulations 1980, and came into force on 1 September 1996. The scope of wastes defined as special was extended by the 1996 Regulations. The new definition of a special waste can be broadly summarised as follows:

- a waste on the EC Hazardous Waste List displaying hazardous properties

- any other controlled waste displaying defined properties, such as flammable, toxic or irritant

- waste prescription only medicines

6.8 Additional wastes such as oils and some photographic chemicals were deemed special wastes for the first time under the 1996 Regulations. Consequently, some businesses became subject to these controls for the first time.

6.9 The Hazardous Waste List is currently under review. A number of changes to the wastes defined as hazardous were agreed in December 1999 and these will need to be implemented into UK law by 1 January 2002. Further changes are being discussed and are likely to be agreed by the end of 2000. As part of that process, the Government has proposed that a number of wastes, most of which are considered special under domestic law, should be added to the European Hazardous Waste List. Wastes that are included on the List will also generally become subject to the Hazardous Waste Incineration Directive and the hazardous waste provisions of the Landfill Directive.

6.10 The Government published in 1999 a consultation document proposing amendments to the existing waste management control system, one effect of which would be to apply the Special Waste Regulations to a wider range of hazardous wastes. This would bring hazardous waste arising from agriculture, mines and quarries within the Special Waste system. A similar consultation paper will be issued in Wales by the National Assembly.

Data on hazardous waste

6.11 The consignment note system produces data on the type, quantity and origin of hazardous wastes to be recovered or disposed of. A register of hazardous (special) wastes moved from a site must be kept for three years. The system is monitored by the Environment Agency and data fed into the Special Waste Tracking System (SWAT).

Special waste arisings 1986/87 – 1997/98[1]

Year	England	Wales	Total
1986/1987[2]	1,500,000	83,000	1,583,000
1987/1988[2]	2,070,000	71,000	2,141,000
1988/1989[2]	1,762,000	60,000	1,822,000
1989/1990[2]	2,146,000	79,000	2,225,000
1990/1991[2]	2,733,000	89,000	2,822,000
1991/1992[2]	2,728,000	122,000	2,850,000
1992/1993[2]	2,216,000	126,000	2,342,000
1993/1994[2]	1,827,000	130,000	1,957,000
1994/1995[2]	1,939,000	133,000	2,072,000
1995/1996[2]	2,339,000	122,000	2,461,000
1996/1997[3]	2,133,000	106,000	2,239,000
1997/1998[4]	4,493,000	385,000	4,878,000
1998/1999[4]	4,452,000	394,000	4,846,000

[1]Does not include special waste treated or disposed of at point of production or hazardous wastes imported under transfrontier shipment regulations. Figures are estimates

[2]Special wastes as defined by the Control of Pollution (Special Wastes) Regulations 1980

[3]Figures are provisional as there is some under-reporting due to the implementation of new regulations, which gave rise to difficulties in collation of the data

[4]Provisional figure. Double counting of waste forwarded to short term storage and waste forwarded for treatment may occur with subsequent forwarding to other treatment/disposal options. The Special Waste Regulations 1996 resulted in additional waste streams being considered for the first time. The figure includes imports from and exports to Scotland. Source: Environment Agency

6.12   Information on special waste arisings up to 1993/94 was calculated by the Waste Regulation Authorities from the consignment notes that accompany and monitor each movement of these wastes. These are estimated figures due to:

- difficulties in reconciling actual amounts produced in, or transferred between, Waste Regulation Authorities with the amounts pre-notified to Waste Regulation Authorities before production and transfer

- difficulties in converting to weight in tonnes from volumetric and other measures used on consignment notes

6.13   The 1996 Regulations resulted in additional waste streams being considered for the first time (such as oils, which alone account for an additional 1,125,660 tonnes in 1997-98).

Special waste arisings by European Waste Catalogue classification

Chapter	Description	Quantity 1997/8 tonnes	Quantity 1998/9 tonnes
Chapter 01	Waste resulting from exploration, mining, dressing and further treatment of minerals and quarrying	6,166	13,446
Chapter 02	Waste from agricultural, horticultural, hunting, fishing and primary production, food preparation and processing	5,517	6,240
Chapter 03	Wastes from wood processing and the production of paper, cardboard, pulp, panels and furniture	11,116	3,313
Chapter 04	Wastes from the leather and textile industry	3,396	1,773
Chapter 05	Wastes from petroleum refining, natural gas purification and pyrolytic treatment of coal	419,452	258,536
Chapter 06	Wastes from inorganic chemical processes	281,056	365,128
Chapter 07	Wastes from organic chemical processes	580,015	572,080
Chapter 08	Wastes from the manufacture, formulation, supply and use (MFSU) of coatings (paints, varnishes and vitreous enamels), adhesive, sealants and printing inks	153,915	145,121
Chapter 09	Wastes from the photographic industry	5,992	12,913
Chapter 10	Inorganic wastes from thermal processes	132,173	107,551
Chapter 11	Inorganic waste with metals from metal treatment and the coating of metals; non-ferrous hydro-metallurgy	174,060	145,927
Chapter 12	Wastes from shaping and surface treatment of metals and plastics	148,084	150,544
Chapter 13	Oil wastes (except edible oils, 05 00 00 and 12 00 00)	860,514	1,030,858
Chapter 14	Wastes from organic substances employed as solvents (except 07 00 00 and 08 00 00)	74,705	108,638
Chapter 15	Packaging; absorbents, wiping cloths, filter materials and protective clothing not otherwise specified	67,487	47,985
Chapter 16	Waste not otherwise specified in the catalogue	261,760	461,727
Chapter 17	Construction and demolition waste (including road construction)	1,259,572	1,007,098
Chapter 18	Wastes from human or animal health care and/or related research (excluding kitchen and restaurant wastes which do not arise from immediate health care)	19,528	16,460
Chapter 19	Waste from waste treatment facilities, off site waste water treatment plants and the water industry	286,728	227,883
Chapter 20	Municipal wastes and similar commercial and institutional wastes including separately collected fractions	64,626	64,575
Chapter 99	Wastes not classified in the European Waste Catalogue	61,788	98,578

General legislative and technical requirements

6.14   The Environment Agency enforces the control system for special waste. Anyone who handles such waste is subject to the Duty of Care under the waste management licensing system, except that the consignment note used to track movements replaces the Duty of Care transfer note. Conviction under the Special Waste Regulations may result in fines up to £5,000 before Magistrates and an unlimited fine at Crown Court and/or imprisonment for up to two years. Further legislative and technical requirements for the management of hazardous waste stem from the instruments listed in the box below.

Other legislation which influences the management of hazardous wastes includes:

- Special Waste (Amendment) Regulations 1997, SI 251
- Environmental Protection (Duty of Care) Regulations 1991 SI 2839
- Waste Management Licensing Regulations 1994 SI 1056, as amended
- Chemicals (Hazard Information and Packaging for Supply) Regulations 1996
- Directives on the disposal of waste oil (75/439/EEC and 87/101/EEC))
- Directive on the disposal on PCBs & PCTs (96/59/EC)
- Directive on the incineration of hazardous waste (94/67/EEC)
- Directive on the Landfill of Waste
- Waste Shipments Regulation 1993
- Pollution, Prevention and Control Regulations 2000
- A number of documents providing guidance on special wastes are available from the Environment Agency and the DETR. These are indicated below.
    - Special Wastes: A technical guidance note on their definition and classification. Joint Environment Agency, Scottish Environment Protection Service and the Environment and Heritage Service, Northern Ireland.
    - Special Waste Explanatory Notes. Environment Agency policy statements on many different aspects of special waste management.

6.15   The DETR/Welsh Office leaflet, *The Special Waste Regulations 1996 – how they affect you*, provides basic information regarding definitions and procedures relating to the 1996 Regulations.

6.16   At present most movements of special waste attract a fee of £15 payable to the Environment Agency. The fee is designed to cover the cost of administering the special waste tracking system in accordance with the *polluter pays* principle. The cost of consigning such wastes in the UK currently stands at around £8 million pounds each year. Suggestions that lower fees should be applied to wastes moving for recovery rather than disposal are being addressed as part of the formal policy evaluation of the Special Waste Regulations.

Hazardous waste reduction

6.17   Reducing the amount of hazardous waste produced is a key waste management priority. By preventing waste from being generated in the first place, environmental burdens associated with resource consumption and waste treatment and disposal are avoided, together with the related financial costs. Where waste production cannot be avoided it may be possible to change the process so that the wastes generated are less hazardous or even non-hazardous.

6.18   A number of issues have increased the awareness and practice of waste reduction in industry, including: new legislation, for example the Special Waste Regulations; the European Integrated Pollution Prevention and Control (IPPC) Directive; increases in the costs of treatment and disposal of hazardous wastes; and increasing awareness of the financial and environmental benefits of reducing both the volume and hazardousness of such wastes.

6.19   The Government's Environmental Technology Best Practice Programme (ETBPP) has also played an important role in promoting cleaner technologies which help reduce hazardous waste arisings.

Examples of ETBPP publications dealing with cleaner technologies relevant to hazardous waste reduction	
Reference	**Title**
GG100	Solvent capture and recovery in practice: Industry examples
GC85	Simple measures to reduce Isopropyl use – Beacon Press
GG199	Optimising the use of metal working fluids
EG39	Perchloroethylene consumption in the dry cleaning industry
GG62	Water and chemical use in the textile dying and finishing industry

6.20   The Government is currently stepping up efforts to encourage hazardous waste reduction. The draft guidelines for Company Reporting on Waste emphasise its importance and challenge companies that produce such wastes to set targets for reductions in quantity and/or hazardousness. Furthermore, guidance on IPPC will set out proposals on waste reduction for the substantial number of industrial and waste management facilities covered by the permitting process. Such steps take place against the background of possible regulatory requirements which may be proposed by the European Commission, notably on extended producer responsibility for electrical and electronic goods and batteries.

Re-use, recovery and recycling

6.21   Re-use, recovery and recycling might not be appropriate for the more hazardous wastes. For example, chemicals that have been banned, such as certain pesticides, PCBs and CFCs, should clearly not be reused, but must be destroyed or treated to remove their hazardous properties. There is, however, some scope for making greater use of these management options for some less hazardous wastes. This is addressed in the environmental reporting initiative and certain projects in the Environmental Technology Best Practice Programme (Chapter 4 section 4.12 of this part of the strategy). Such work will be supplemented by projects under the new sustainable waste management programme, including one on the regeneration of waste oil.

Examples of hazardous waste re-use, recycling and reclamation	
Option	**Suitable waste streams**
Recovery for use as fuel	Organic solvents may be blended to produce secondary liquid fuels (SLF) which can be used as a substitute for fossil fuels
Solvent reclamation/recovery	Organic solvents including halogenated solvents, phenols, ethers etc can be regenerated, for example by re-distillation
Recovery of precious metals	Recovery of silver from photographic chemicals, regeneration of spent catalysts
Recovery of fine chemicals	Fine chemicals, biocides, organic compounds
Re-refining of waste oils	Waste mineral oils such as engine oils
Recovery of heavy metals	Car batteries, nickel cadmium batteries, recovery of mercury from fluorescent light tubes

6.22 The direct reuse of hazardous waste materials without reprocessing, either on or off the site of generation, is influenced by a number of issues including:

- perceived and actual quality of the waste material
- consistency of quality and volume of the material
- the relative value of secondary materials and virgin materials
- perceptions of the use of waste
- level of awareness of the potential of the reuse of wastes
- degree of hazardousness of the waste

Hazardous waste incineration

6.23 The Government is committed to the principles of self-sufficiency and proximity in the disposal of wastes. It recognises the importance of having available a network of high temperature incinerators, suitable for the disposal of hazardous organic wastes and other wastes where high temperature incineration is the Best Practicable Environmental Option. These facilities must be subject to high standards of protection for human health and the environment and should be fitted with energy recovery, where practicable.

6.24 As discussed in the consultation paper on the UK Management Plan for Imports and Exports of Wastes, such facilities also enable the UK to contribute towards the resolution of global environmental problems by disposing of hazardous wastes from countries which are not yet in a position to build and maintain specialised facilities of this sort. The consultation paper also considers whether the UK should continue to accept imports of hazardous waste for high temperature incineration from the Republic of Ireland.

6.25 The Hazardous Waste Incineration Directive (94/67/EC) was adopted in 1994 and sets out stringent environmental performance standards for hazardous waste incinerators. New facilities within the scope of the Directive are required to comply with its requirements immediately. Existing facilities are required to comply by the end of June 2000. The Hazardous Waste Incineration Directive is being merged with the proposed Waste Incineration Directive, but it is likely to remain in force until around 2005.

6.26 The Hazardous Waste Incineration Directive applies to all incinerators burning waste on the EC Hazardous Waste List (subject to specified exemptions, for example certain waste oils). Such incinerators need to be authorised by the Environment Agency, under the integrated pollution control system. Facilities for which authorisations are required include:

- merchant chemical waste incinerators (for example those which receive waste from other waste producers)

- in-house chemical waste incinerators

- some drum reconditioners

- cement and lime kilns (which co-incinerate hazardous waste oils and solvents which are not otherwise excluded)

6.27 Hazardous wastes for which high temperature incineration is generally considered the most suitable disposal option on environmental and safety grounds are also often highly combustible. Examples of such wastes include: agrochemical residues; halogenated wastes; oils containing polychlorinated biphenol; solvents; laboratory chemicals; and acid tars. However, high temperature incineration operators increasingly favour the mixing of such wastes with aqueous wastes, including other hazardous wastes, to create blends suitable for efficient burning.

6.28 There are three merchant hazardous waste incinerators in England and Wales. These high temperature incinerators have a combined capacity of 134,000 tonnes per annum. The Government and the National Assembly are currently studying the likely impact of the Landfill Directive on the pattern of hazardous waste disposal, including whether current high temperature incineration capacity would be sufficient to deal with any diverted waste.

Merchant hazardous waste incinerators in England and Wales			
Operator	**Location**	**Date commissioned**	**Design capacity (tonnes per annum)**
Cleanaway	Ellesmere Port	1990	75,000
ReChem	Southampton	1990	35,000
Shanks Chemical Services Ltd	Pontypool	1972	24,000

6.29 In studying this question, the Government and the National Assembly will consider whether any additional high temperature incineration capacity could readily become available. The existing merchant incinerator at Southampton has already obtained planning approval for an extension; or at any other facilities, consistent with the generally more rigorous controls that will apply with the introduction of the IPPC regime.

6.30 Any proposals for extra high temperature incineration facilities to be constructed at existing merchant incinerators, or at new sites would have to be considered through the normal processes of the land use planning system, taking account of the requirement for a network of suitable facilities for hazardous waste disposal. The Environment Impact Assessment Regulations 1999 require that any planning application for an installation for the incineration, chemical treatment or landfill of hazardous waste must include an Environmental Impact Assessment.

6.31   In addition to the merchant hazardous waste incinerators, there are fourteen in-house incinerators treating materials such as resin containing liquids, industrial gases, explosives, metal containing wastes, volatile organic compounds containing liquids, catalysts, contaminated soil, and various wastes from the oil and pharmaceutical industries. These incinerators have design capacities between 0.15 tonnes per hour to 22.5 tonnes per hour.

6.32   Four drum reconditioning facilities include incineration as part of their operations. The primary aim of the operations is to remove hazardous materials from drums through incineration prior to reshaping, painting and finishing the drums. Facilities are currently located in Aldershot, Avonmouth, Barking and Deeside.

6.33   Conventional fuels are increasingly being substituted by alternative fuels in cement and lime kilns. The dominant alternative fuel is secondary liquid fuel which is derived from selected waste solvents and supplied to a specification. There are currently nine cement and lime kilns authorised to either burn or run trials with secondary liquid fuel.

Cement and lime kilns burning secondary liquid fuels		
Site	**Kiln**	**Company**
Ketton	Cement (2)	Castle
Clitheroe	Cement (2)	Castle
Barrington	Cement	Rugby
Weardale	Cement	Blue Circle
Masons	Cement	Blue Circle
Whitewell	Lime	Redland
Thrislington	Lime	Lafarge

6.34   The Government and the National Assembly attach priority to the regeneration of waste oils for re-use as lubricating oil, in line with the Waste Oils Directive, and oils should be regenerated whenever it is possible to do so. However, they recognise that the next best option is generally for it to be burned as fuel and that this is generally the Best Practicable Environmental Option for waste oils that are not suitable for regeneration. Approximately 350,000 tonnes per annum of waste oil is burned. Such oil is hazardous, but is generally exempted from the provisions of the Hazardous Waste Directive, though not those of the proposed Waste Incineration Directive, which is likely to apply from about 2005.

Landfill

6.35   The recently adopted Landfill Directive will impact significantly on the disposal of hazardous waste, as it introduces more stringent controls in order to protect human health and the environment. In 1997/8 and 1998/9, 53% and 47% of hazardous wastes respectively were disposed of to landfill. The implications of the Directive include:

- an end to the practice of co-disposal (for instance where hazardous and non-hazardous wastes are disposed of in the same landfill)

- landfilling of hazardous materials only in facilities specifically designed to accept the materials

- increased requirement for pre-treatment of waste prior to landfilling

- banning of certain materials from disposal to landfill, if they possess for example, corrosive, oxidising, flammable, or liquid properties

6.36 Other aspects of the Directive that will affect the disposal of hazardous waste to landfill include meeting the costs of a minimum of 30 years aftercare following closure and meeting criteria regarding acceptance procedures and monitoring regimes.

6.37 Hazardous wastes which have previously been considered unsuitable for landfill disposal in the UK, because of their intrinsic properties and the risk they pose, include:

- acid tars
- flammable wastes
- many organic solvents
- explosive wastes
- waste which react violently with water or organic matter
- wastes containing significant concentrations of certain organic compounds (such as polychlorinated biphenyls (PCBs), and polychlorinated triphenyls (PCTs))

6.38 Three types of landfill are currently used for the disposal of hazardous waste – co-disposal, monodisposal and multidisposal. Co-disposal is the disposal of hazardous wastes with household or other similar waste. The process utilises the properties of household-type waste to attenuate those constituents in hazardous wastes which are polluting and potentially hazardous and thereby make their impact on the environment acceptable. A balanced input of hazardous and household waste is required to ensure that the attenuation processes are not overwhelmed. The practice requires special precautions and management of all operations to ensure that it is both safe and environmentally acceptable. Those wastes destined for co-disposal are assessed to ensure compatibility with household waste. Under the Landfill Directive, co-disposal will no longer be permitted for new sites, and will be phased out for existing sites.

6.39 Monodisposal is the disposal of wastes having the same general physical or chemical form by landfill or lagooning. Following disposal, the waste need not necessarily remain in the same physical form as it was produced. For example, pulverised fuel ash from power stations is almost always landfilled at monodisposal sites and frequently it is pumped there as a slurry and allowed to dry out. Producers of bulk inorganic chemicals often dispose of large quantities of waste at monodisposal landfills or lagoons.

6.40 Multidisposal is generally used to describe the disposal of chemically different wastes that, like liquids or sludges, have similar physical forms. The deposit of mixed wastes either as liquids into lagoons or at sites accepting both inert and degradable industrial and commercial solid wastes may also be regarded as examples of multidisposal operations. The Landfill Directive will ban the disposal of any liquids to landfill.

Pre-treatment

6.41 Pre-treatment can reduce the hazardousness of waste or, in some cases, render it non-hazardous. Under the Landfill Directive, pre-treatment will be necessary for most hazardous wastes which are to be landfilled. Certain treatments may be used which do not alter the hazardous properties of a waste but significantly reduce the probability of the hazard having an effect. A wide variety of pre-treatment techniques are available, with biological, thermal and some physico-chemical treatments most suitable for organic wastes and physico-chemical treatment most suitable for inorganic wastes. Newly emerging energy recovery technologies can be expected to become increasingly prominent, encouraged by the need to divert waste away from landfill. Some of these new processes, notably plasma arc (see Chapter 5 section 5.80 of this part of the strategy), will be suitable for hazardous waste streams.

6.42 The implementation of Directives regarding landfill and the incineration of hazardous waste will significantly influence the environmental, technical and economic framework within which options for the management of hazardous waste sit. As a result of such changes, new waste management options will need to be developed relating to both the pre-treatment of hazardous waste and its disposal. These issues are being addressed in the study into the impacts of the Landfill Directive on hazardous waste management in the UK, which will assist the Government and the National Assembly in further developing their strategies on hazardous waste matters.

Issues for the BPEO for hazardous waste

6.43 DETR and the Environment Agency are developing proposals on how best to proceed with formal assessments of the Best Practicable Environmental Option (BPEO) for key hazardous wastes streams. The determination of BPEO for the management of hazardous wastes will need to be carried out on a waste by waste basis in the context of available waste treatment facilities, using techniques such as life cycle assessment.

6.44 The development of BPEO for the future management of hazardous waste will need to consider a number of issues. These will be based around the waste hierarchy, though the hierarchy will not always apply to hazardous wastes in the same way that it would to non-hazardous wastes. Issues will include:

- re-use, recycling or reclamation of waste either at the site of generation or elsewhere. Re-use and recycling will not be appropriate for all hazardous wastes. Banned substances should not be re-used when they arise as waste and recycling should not result in the *spreading* of contaminants for example asbestos should be removed from feedstock for the crushing of demolition waste to prevent it spreading amongst recycled aggregates

- reclamation of energy from waste

- incineration without energy recovery. This may remain appropriate for certain waste streams such as PCBs, CFCs, pesticides and halogenated and non-halogenated solvents

- constraints on landfill (including the implications of the landfill Directive). Landfill may remain the appropriate option for some waste streams such as asbestos and some treated timber

- specialised treatment to reduce hazardous properties even if this results in an increase in the quantity of waste

- the proximity principle

- environmental receptors sensitive to the waste

- existing and alternative management practices

6.45   To assist in developing priorities, it will be necessary for this work to be co-ordinated with future assessments of BPEO for industrial processes, where the objective is to reduce the generation of waste at the point of production. This would include both quantitative reductions to avoid the need for treatment and disposal and qualitative reduction to reduce the environmental impact of such operations (e.g. removal of mercury from consumer batteries will make the recycling of separately collected batteries much easier).

6.46   Chapter 8 of this part of the strategy includes more detailed material on special arrangements for particular waste streams, including hazardous ones. The Government and the National Assembly intend to develop this analysis through further work on these and other waste streams.

CHAPTER 7

Dealing with packaging and packaging waste

7.1 Packaging waste is the only element of the waste stream currently subject to producer responsibility regulation. Packaging is about 9% of industrial, commercial and municipal waste and is currently estimated at around 10 million tonnes annually, of which around 4.5 million tonnes is packaging waste arising in the household waste stream.

7.2 Packaging can reduce product wastage and save resources between the point of production and final consumption. Nevertheless, recovery and recycling of packaging waste can contribute to the commitment of the Government and the National Assembly to sustainable development and tough climate change targets. Positive benefits include:

- reducing biodegradable packaging going to landfill
- reducing methane emissions
- achieving energy savings
- limiting the depletion of natural resources
- decreasing other environmental impacts

7.3 To achieve optimal reduction in overall waste, continued emphasis will need to be placed on addressing the management of packaging and packaging waste.

EC Packaging Directive obligations

7.4 The UK is committed to meet the objectives and targets in the EC Directive on Packaging and Packaging Waste. The first priority is to reduce the amount of packaging waste for final disposal. The Directive also requires Member States to achieve certain packaging waste recovery and recycling targets, and ensure that packaging meets certain essential requirements. The targets, to be met by end June 2001, are:

- to recover between 50% and 65% of packaging waste
- to recycle between 25% and 45% of packaging waste
- to recycle at least 15% of each material

7.5 The European Commission is expected to propose new targets for the period 2001 to 2006 in the course of 2000. Member States have discretion to implement the Directive as they see fit. In the UK, because recovery will have to rise from around 30% in 1997 to 50% in 2001, interim targets have been set so as to facilitate planning for 2001 targets.

Performance reported by other Member States on packaging*					
Country	Waste stream	Year	Recovered	Of which recycled
Germany	household, commercial and industrial	1997	92%	86%
Netherlands	household	1996	73%	52%
France	household	1997	45%	29%
Italy	household, commercial and industrial	1996	34%	31%
UK	household, commercial and industrial	1998	33%	29%
Spain	household	1997	31%	21%

*Caution should be exercised in interpreting these figures, since calculation methods vary from country to country

UK response to the EC Directive

7.6 The UK has implemented the Directive through the Producer Responsibility Obligations (Packaging Waste) Regulations 1997, amended by two further Statutory Instruments in 1999[1]; and by the Packaging (Essential Requirements) Regulations 1998. The Producer Responsibility Regulations came into force in March 1997 following extensive consultation with industry both on the form that the obligation to recover and recycle packaging waste should take on the draft Regulations. A review of the Regulations was conducted in 1998 and resulted in a series of amendments, which were also the subject of consultation with the industry. Some of these came into force in 1999 others in 2000. The main amendments were:

- targets up in 1999 from 38% recovery, 7% recycling to 43% recovery, 10% recycling; and in 2000 to 45% recovery, 13% recycling

- new consumer information obligations

- simplification of the system (including removal of the 'wholesaler obligation' and the competition scrutiny regime for compliance schemes

7.7 The packaging (Essential Requirements) Regulations 1998 took full effect in January 1999. Authority for their enforcement is vested in Trading Standards Officers. They implement the Directive provisions specifying essential requirements for packaging placed on the market, which cover minimisation, avoidance of noxious and hazardous substances and the need for packaging to be recoverable (through recycling, energy recovery or composting).

7.8 The Packaging Regulations place four main obligations on businesses (producers) which satisfy two threshold tests. The obligations are:

[1] The Producer Responsibility Obligations (Packaging Waste)(Amendment) Regulations 1999 SI No 1361, and the Producer Responsibility Obligations (Packaging Waste)(Amendment)(No.2) Regulations 1999 SI No. 3447

- to register with the Environment Agency and provide specified packaging data

- to recover specified tonnages of packaging waste, the tonnage calculation being based on the weight of packaging handled by the business, the relevant target and the packaging activity being performed

- to certify annually that the obligations have been met

- to inform consumers about their role in increasing recovery and recycling (retailers only)

7.9 Obligated producers may choose whether to meet their obligations themselves and register individually with the Environment Agency, or join a compliance scheme. By the end of 1999, there were 4,223 registrations representing 9,500 businesses. Around 80% of registrations were with compliance schemes of which 13 are registered.

7.10 On registration, producers are required to provide the Agency with data on the packaging handled by their businesses. This information forms the basis for:

- calculating individual recycling and recovery obligations

- assessing the total weight of packaging waste and the UK's progress towards the EC Directive targets

- reporting UK packaging information to the European Commission

7.11 There have been difficulties in collecting robust data from a sector as diverse as the packaging chain and there were significant differences between the packaging reported in 1997 and 1998. However, data continues to improve and the development of systems to procure accurate data will continue to be given high priority.

7.12 The Environment Agency provides guidance to businesses on, for example, what is and is not packaging. It has a duty to monitor compliance with the Regulations in England and Wales and, where appropriate, to take enforcement action. The Agency also operates a voluntary accreditation scheme for reprocessors who, once accredited, issue evidence of compliance with recovery and recycling obligations (for example Packaging Waste Recovery Notes (PRNs)). Such evidence is used by obligated producers as evidence that recovery of packaging waste has been undertaken. 207 reprocessors were accredited in 1998 and 210 in 1999.

Forecasts – packaging waste and recovery levels

7.13 A major expansion of collection and recovery capacity is needed to ensure that the UK can keep in step with European target levels of packaging waste recovery. Reduction of packaging and the development of end-use markets for recycled materials will continue to be key. It is difficult to predict the growth of packaging placed on the market, as this depends on social, demographic and economic circumstances, but a sustained effort will be needed to continue to reduce packaging and packaging waste.

1998 packaging waste recovery achieved (tonnes)				
Packaging in the waste stream (estimated) all materials	Recovery achieved 1998	Recycling	Recovery (%)	Recycling (%)
10,000,000	3,338,705	2,720,351	33%	27%

Represents consolidation of returns made by the Environment Agency and SEPA for England, Wales and Scotland and includes an estimate of wood recovery (170,000t)

7.14 The UK met its own expectations for recycling and recovery for 1998 and already, with 27% recycling, meets the Directive 25% target for minimum recycling overall 3 years ahead of target. The material-specific recycling targets for each material (15% minimum in 2001) have also already been met in 1998 for all materials except aluminium (13%) and plastic (8%). Plastics still has the longest way to go, and overall recovery will have to rise by 17% between 1 January 1999 and 30 June 2001 to meet the Directive targets.

Critical factors in achieving targets

7.15 Packaging waste arises in both the commercial/industrial and household waste streams. Around half of all packaging placed on the market ends up in the household waste stream. Some materials (for example glass or aluminium) are found principally in the household stream. The collection of some material is economically viable now, but for the Directive targets to be met, appropriate levels of funding will have to be dedicated to building up collection systems, particularly for household waste, and the reprocessing infrastructure; and also to developing end-use markets for recycled materials. Local authority and other collection systems can only operate economically if there are satisfactory market values for recycled materials. A sound and growing infrastructure of collection systems (particularly for steel, aluminium, plastic and glass packaging waste) will be a prerequisite to meeting recycling and recovery targets as will the development of sufficient markets for recycled materials.

Estimated proportion of packaging materials arising in the household and commercial/industrial waste streams)				
Material	Arising in the household stream		Arising in the commercial/ industrial stream	
	Tonnes	%	Tonnes	%
Aluminium	104,500	96%	4,500	4%
Steel	573,000	78%	162,000	22%
Plastic	1,100,000	71%	600,000	39%
Glass	1,850,000	84%	350,000	16%
Paper	500,000	13%	3,470,000	87%

7.16   The capacity to reprocess sufficient packaging waste to meet targets in 2001 and beyond varies greatly from material to material. Estimates have been made using the best information available, but these are affected by a degree of uncertainty, particularly in the extent and quality of packaging waste data. Moreover, external factors – virgin material prices and global waste material prices, which affect the demand for waste materials in the UK – will influence further investment in recycling plant.

Packaging recovery rates (GB)		
Material	Achieved 1998	2001 Directive recycling & recovery targets
Paper	1,894,086	
Glass	503,800	
Steel	182,409	
Aluminium	14,517	
Plastic	125,539	
Wood/other		
Total recycling	2,720,351	2,500,000 (25%)
Energy from waste/refuse derived fuels/Composting	448,354	
Wood recovery	170,000	
Total recovery	3,338,705	5,000,000 (50%)

Material specific obstacles to increased reprocessing

7.17   The development of new, and the expansion of existing, end-use markets is needed for all materials, but different materials have particular problems:

- *glass*: the major obstacle is the over-supply of green cullet arising from excessive imports in green glass (principally wine and beer) and the absence of UK markets for this material

- *steel*: the challenge will continue to be to increase collection, particularly from the household waste stream. The industry has declared that, at the moment, there is no practical limit to the amount of steel packaging waste it can recycle

- *aluminium*: total capacity to reprocess aluminium scrap in the UK is 375,000 tonnes. Increasing collection is critical and the rate of collection will determine how much of this capacity will be utilised

- *plastics*: the lack of capacity to collect and recycle plastic packaging waste remains the most significant obstacle to the UK's overall reprocessing ability. Current capacity to recycle plastic packaging waste is 150,000 tonnes, rising to 255,000 tonnes in 2001. If this level is achieved the UK is likely to meet the minimum recycling requirement (15%) under the Packaging Directive, but this will only contribute marginally to the much larger recovery target. Collection from the household waste stream continues to be a challenge for this material too

- *wood and other*: the obligation to recover packaging manufactured from wood and other materials (ceramics, jute, hessian and the like) took effect on 1 January 2000, so there is little experience on which to base predictions of achievement. Wood packaging is likely to make up a proportion of the waste going to composting and energy from waste plants. In 1998, although there was no obligation to recover or recycle wood packaging, some 170,000 tonnes of wood packaging waste was recovered

- *energy from waste*: energy-from-waste plant can now absorb 2.5 million tonnes of mixed waste, including 475,000 tonnes of packaging waste, mostly plastics and paper. New plant under construction will add a further 120,000 tonnes to this capacity, but it should be noted that only half of this will come on stream by 2001

- *composting*: some packaging waste is composted but the amounts are insignificant

- *minimisation*: reduction of packaging has been ongoing for some years but there has also been a trend towards packaging a greater number of products and to added reliance on packaging to provide greater product protection. The increase in the amount of packaging also reflects demographic influences, particularly the reduction in the average number of inhabitants per household. Continued reduction of packaging and packaging waste is needed while continuing to use packaging in the interests of safety, hygiene and consumer acceptance. Greater consideration also needs to be given to re-using packaging where practicable

Consumer awareness and contribution

7.18   The Packaging Directive obligates Member States to ensure that users of packaging, including in particular consumers, are informed about the role which they play in contributing to re-use of packaging and the recovery and recycling of packaging waste and the means by which these activities are performed. The Secretary of State has responsibility to ensure that users of packaging are so informed, but the Regulations also place certain responsibilities on retailers to increase public awareness of, and participation in, recycling activities to help meet UK targets. DETR also provides a variety of information about packaging and recycling in its publication *A Forward Look for Planning Purposes*, and there is a non-statutory Guidance Note on the Regulations called *The User's Guide*.

CHAPTER 8

Progress with various waste streams

8.1 This chapter of the waste strategy is devoted to considering a number of specific waste streams. This list is by no means definitive, but does cover most of the significant waste streams, or specific waste streams which have been selected for action at the European or UK level. Note that hazardous and special wastes are covered in Chapter 6, and packaging wastes are dealt with in Chapter 7 of this part of the strategy. Please note that unless otherwise stated, figures quoted in this Chapter are taken from the *DETR Digest of Environmental Statistics*, or have been supplied by the Environment Agency.

Agricultural wastes and pesticides

8.2 Much of the waste and by-products arising on farms consists of organic matter such as manure, slurry, silage effluent and crop residues. 80 million tonnes of by-products and waste arise annually from housed livestock alone. Besides these wastes, the next significant wastes are packaging and films, and sheep dip.

8.3 At present, waste from premises used for agriculture is excluded from the definition of controlled waste, and hence is not subject to the waste management licensing regulations, or other waste controls such as the Duty of Care and registration of carriers. These controls are to be extended to agricultural waste. Consultation papers setting out proposals will be issued by the Government and the National Assembly for Wales in 2000. We are keen to ensure that these controls are proportionate and impose minimum burdens on the industry, consistent with fulfilling our European obligations.

8.4 As a consequence of extending waste management controls to agricultural waste it will be necessary for farmers to apply to the Environment Agency for a waste management licence to carry out operations which are not exempt from licensing. There are legal obligations and costs associated with holding a licence and farmers may find it easier and cheaper to send waste to someone else for recovery or disposal at a licensed site rather than manage it themselves.

8.5 The Government has recently issued revised *Codes of Good Agricultural Practice for the Protection of Water, Soil and Air* which provide practical guidance and advice on the storage, management and application of a wide range of farm by-products and wastes.

8.6 When agricultural wastes are classified as controlled wastes, there will be an obligation on the part of the waste holder (the farmer) to decide whether the waste should be treated as special waste under the *Special Waste Regulations 1996*. Other wastes from agriculture that may be classified as special waste include cement asbestos from farm buildings and oil and batteries from farm vehicles and machinery. The deposit, recovery or disposal of agricultural waste will require a waste management licence once these changes have been implemented.

However, where disposal is already subject to a permit under the Groundwater Regulations 1998, arrangements will be made to ensure that dual control is not applied.

Likely production of various categories of special waste from agricultural sources:	
Special Agricultural Waste	**Annual production**
Agrochemicals packaging	3,500 tonnes
Waste oil	25,000 tonnes
Batteries	500 tonnes
Sheep dip	70,000 cubic metres

REDUCING AGRICULTURAL WASTES

8.7   There are considerable opportunities for reducing and recovering agricultural wastes. In 1996 and 1997 research projects, sponsored jointly by the Ministry of Agriculture, Fisheries and Food (MAFF) and the BOC (British Oxygen Company) Foundation, were set up to investigate whether industrial waste reduction principles could be applied in agriculture. Information was collected on existing practices and disposal routes. Savings were identified in a wide range of waste streams including wasted animal feeds, water and energy consumption, purchased fertilisers, waste pesticides, wasted crop produce in the field or store, and packaging.

8.8   As a result of the research, a waste minimisation manual was published in January 2000. The manual is designed to enable farmers to review many of the tasks carried out on their holdings and then to identify waste reduction measures which will both save money and benefit the environment.

SPECIFIC AGRICULTURAL WASTE STREAMS

8.9   The following waste streams often need to be managed separately from other waste streams, either to reduce the threat that the waste's hazardousness represents both to human health and the environment, or to ensure the waste stream is effectively managed.

Farm by-products

8.10   The most economical and environmentally safe way of managing farm by-products such as manures and slurry is to obtain value from them by properly applying them to land. When applied to land they provide valuable nutrients and organic matter and allow farmers to reduce the amount of inorganic fertiliser applied.

8.11   This type of organic matter when used in this way will fall outside the scope of the legal definition of waste, and may be described as an agricultural by-product. It will not therefore be subject to the controls on waste in Part II of the Environmental Protection Act 1990, and other waste legislation.

Farm films

8.12   The collection of farm films, or non-packaging farm plastics, has been a particular agricultural waste problem, not least because they are bulky, often heavily contaminated with soil and expensive to dispose of off-farm. Therefore, farmers have tended to bury or burn these wastes. The difficulties and expense of disposing of farm plastics will become more significant when waste management controls are extended to agricultural waste, thereby effectively halting on-farm disposal of these plastics. Following the collapse in early 1997 of the Farm Film Producers Group, a voluntary collection and recycling scheme, the Government considered how to approach the issue and a consultation document was published in 1998.

8.13   The Government consulted on two options for the collection of farm films. The first option, and the one favoured by the Government, will encourage a voluntary approach, but this time with the forthcoming extension of waste management controls to the agricultural sector acting as the driving force. The second option is to introduce producer responsibility regulations that could place statutory obligations on plastics manufacturers to recover farm plastics.

8.14   In Wales, the National Assembly is supporting an Objective 5b EAGGF project – Second Life Plastics Wales – to recover agricultural waste plastic in rural Wales with an award of grant under the Rural Development Grant Scheme. The project, which farmers pay a membership fee to join, is also being supported by the Environment Agency, the Welsh National Parks and the Countryside Council for Wales. It is estimated that around 4,000 farmers are signed up to participate, although the project aims to increase this to 7,000 by the end of the three years for which the grant is available.

Pesticides and veterinary medicines

8.15   Farmers need to dispose of pesticides and veterinary medicines properly, and it is important for them to have a clear plan for dealing with such wastes. As they own the waste, it will be the responsibility of farmers under their impending Duty of Care to ensure it is disposed of carefully, or preferably collected by a specialist contractor. When agricultural wastes are brought under the definition of controlled waste, pesticide waste products such as concentrates and unrinsed packaging, and veterinary medicines are likely to be classified as special waste.

8.16   The disposal of used sheep dip and dilute pesticide washings to land requires an authorisation from the Environment Agency under the Groundwater Regulations 1998, which came fully into force on 1 April 1999. The Government will provide guidance on these Regulations. When sheep dip becomes a controlled waste, it is likely to be classified as a special waste. Its disposal will be subject to requirements for a permit. This may be a Waste Management Licence or a permit issued under the Groundwater Regulations.

Other agricultural waste

8.17   Cement asbestos sheeting has been widely used in the construction of farm buildings. When agricultural waste becomes controlled waste, asbestos from the demolition or repair of these buildings will be classified as special waste and its disposal will require a waste management licence. Used engine oil is also classified as special waste and therefore oil from farm vehicles and machinery similarly will become special waste.

Batteries

8.18   Apart from a number of specialist varieties, batteries can conveniently be classified into two main types: automotive and consumer. The former are normally recycled, but the vast majority of the latter are thrown away, a situation which will have to change in coming years if, as expected, EC legislation sets collection and recycling targets for these wastes for the first time. (Data on battery wastes supplied by Department of Trade and Industry).

AUTOMOTIVE BATTERIES

8.19   Around 10 million lead acid automotive batteries are sold in the UK each year and a similar number are scrapped. Figures for 1997/98 show that 140,000 tonnes of lead acid batteries were consigned for recycling in England and Wales. For 1998/99 this rose to 144,000 tonnes.

8.20   The components of lead acid batteries are separated for recycling purposes. The lead and acid contents of used batteries are classed as hazardous waste, and whole batteries are delivered to a central point for recovery. The lead plates are sent to smelters for re-smelting, the polypropylene cases can be sent for recycling into other polypropylene items such as video casings, and the sulphuric acid can be re-used as low grade acid, regenerated or sent for neutralisation and disposal. There is generally an economic incentive for the recycling of lead acid batteries, due to the value of the lead metal in the battery plates.

8.21   Lead acid batteries are classified as special waste due to the content of sulphuric acid (H8 Corrosive) and Lead Compounds (H10 Teratogenic). Similar lead acid batteries are used for purposes other than motor vehicles such as telephone exchanges and back-up power supplies. These can be considered in the same way as automotive batteries.

8.22   The collection infrastructure has traditionally been pyramidal, with numerous small collectors forming the base. The wide base of small collectors has given rise to difficulties in ensuring compliance with the Special Waste Regulations in some cases. The implementation of the Special Waste Regulations and, in particular, the £10 consignment fee does not appear to have affected collection and recycling rates. There is however some evidence to suggest that the patterns of collection of lead acid batteries is changing and that battery manufacturers, distributors, large scale collectors and large garages are starting to play an increasingly important role.

8.23   It has been argued that the commitment to maintaining a better than 90% recycling rate is being harmed by sharp falls in world lead prices, which undermines the economics of the collection chain. It has been suggested that formalised arrangements should be developed to support recovery at such times. A range of options for implementing a Batteries Directive which requires the achievement of recycling targets will be considered.

CONSUMER BATTERIES

8.24   Consumer batteries can be categorised into single-life types such as zinc-carbon, alkaline-manganese and various button cells, and rechargeable varieties such as nickel-cadmium and nickel-metal hydride. It is estimated that around 600 million consumer batteries are thrown away each year, representing between 20,000 and 40,000 tonnes. Much of this waste arises as household waste. All batteries contain hazardous substances in various quantities. Where consumer batteries are not household wastes, for example batteries for business appliances, then they may be classified as special waste (such as nickel cadmium batteries, or mercury dry cells). At present, fewer than 1000 tonnes of consumer batteries are classified as special waste and consigned annually in England and Wales.

8.25   Some batteries contain valuable materials. Nickel scrap solids had a value of £4,500 per tonne, and zinc of £450 per tonne at the end of 1999. Technologies exist for the treatment and recycling of all battery types. In the Netherlands batteries are collected, and landfilling of batteries is banned. Heavy metals are often recovered in Sweden, France and the United States. Mercury and silver are recovered in the Netherlands, and nickel-cadmium in the Netherlands and France. Silver compounds from batteries can be recovered in the UK. However, consumer battery recycling is restricted by the cost of collecting, segregation and recycling of the materials compared with their value. General purpose battery recycling facilities in Switzerland are believed to charge more than £2,000 per tonne. In the past there were also problems with hazardous components, especially mercury, in batteries which made recovery of other materials technically difficult. There are no recovery facilities for the recycling of common battery types in the England and Wales.

8.26   The battery industry is of the view that recycling will become more commercially viable when mixed battery feedstock becomes mercury free. The EC has recently adopted lower mercury tolerances, which should speed up this process. Clearer labelling of different battery types or standard colour coding might make segregation at source or after collection easier. Waste reduction (by volume and by hazardousness) might be achieved by greater use of rechargeable batteries and/or reducing the power ratings of battery driven appliances.

8.27   Only nickel-cadmium consumer batteries are collected in any numbers. As part of the UK's implementation of the 1991 Batteries Directive, an industry-led programme (REBAT) is recovering increasing quantities from industrial and commercial sources, and these are being shipped to France for recycling. The scheme has been set a target of recycling 1,550 tonnes of battery wastes over four years.

THE FUTURE

8.28   The 1991 Batteries and Accumulators Directive (91/157/EEC) applies only to primary and secondary batteries containing lead, mercury or cadmium (less than 10% of all batteries sold). The European Commission is expected to come forward shortly with a proposal to revise the Directive by extending its scope to cover all batteries, and setting quantified collection and recycling targets.

8.29   In preparation for the publication of the Commission's proposal, the Government and the National Assembly are assessing a range of collection infrastructures and reprocessing options that might have the potential to recover large quantities of spent consumer batteries from industrial, commercial and domestic sources, and are seeking to quantify the costs, benefits and environmental impacts of what the Commission is expected to propose.

Clinical waste

8.30   The chief inherent hazards of clinical waste are pathogens and sharps. The people chiefly at risk are clinicians and medical staff. Waste handlers are at some risk, while members of the general public face little or no risk. All the hazards can be sufficiently managed by well understood procedures of infection control and risk management. These procedures rest crucially on the competence of the control of infection team, and their ability to influence others towards methods that are adequate and effective without being extravagent. Thus the practical arrrangements for managing clinical waste depend on the different hazards they present.

8.31   Clinical waste is defined in the Controlled Waste Regulations 1992 as:

- human and animal tissue, or blood or other bodily fluids, or excretions, and drugs or other pharmaceuticals

- swabs or dressings syringes, needles or other sharp instruments, which unless rendered safe may prove hazardous to any person coming into contact with it

- any other waste arising from medical, nursing, dental, veterinary, pharmaceutical or similar practice, investigation, treatment, care, teaching or research, or the collection of blood for transfusion, being waste which may cause infection to any person coming into contact with it

8.32   Some clinical waste is classified as special waste. These wastes are subject to additional controls under the Special Waste Regulations. Waste prescription-only medicines are special waste. The Environment Agency's Special Waste Explanatory Note *Health Care Waste* (SWEN001) indicates which clinical wastes are to be regulated as special waste.

8.33   Principal sources of clinical wastes include hospitals, nursing homes, health centres, veterinary surgeries, dental surgeries, GP surgeries, blood transfusion centres and public health laboratories. Other sources include medical, dental and veterinary teaching and research establishments.

8.34   Wastes such as sanitary towels, nappies and incontinence pads (known collectively as sanpro waste) are not considered to be clinical waste when they originate with a healthy population.

REDUCING CLINICAL WASTES

8.35   The best available estimate indicates that NHS trusts in the UK produce 193,000 tonnes of clinical waste each year. This estimate takes some account of the disposal – often as the result of rational circulation – of substantial quantities of non-infectious and non-hazardous wastes in the same bags with clinical waste as properly defined. No similar estimate has been made of UK clinical waste from other sources. Some surveys indicate that it might be in the range 100,000–200,000 tonnes per year.

8.36   An Audit Commission report published in 1997 identified a number of opportunities for reducing, re-using and recycling clinical waste. These included:

- increasing the use of re-usable devices – provided they can be sterilised without too much difficulty

- careful categorising and segregation of the waste to ensure that a minimum of household waste is mixed with the clinical waste

8.37   The Audit Commission emphasised the value of waste segregation in controlling the cost of managing clinical waste.

PRE-TREATMENT OF CLINICAL WASTES

8.38   Clinical waste may be treated prior to disposal. Some treatments so alter the state of the waste that it ceases to be clinical waste. Others at least reduce the risk of infection. Thus these techniques increase the range of final disposal options.

8.39   All treatment options use heat, chemicals, irradiation or a combination of these processes, with or without macerating or shredding the waste. Some systems are suitable for use on the site where the waste is generated. The selection of a system depends mainly on the composition and volume of the waste, and the costs of setting up and operating the system. The primary objective of all these systems is to reduce the concentration of pathogens in the waste. Many also diminish or remove the offensive characteristics of healthcare waste:

- *Heat treatment* – Some thermal systems use relatively low temperatures (150°C – 250°C); others use high temperatures (approximately 500°C to greater than 6,000°C). Low temperature systems make use of wet heat and pressure (autoclaves), dry heat, microwaves or macrowaves. High temperature systems include pyrolysis, plasma technology and gasification, although these processes have not been generally adopted for clinical waste.

- *Chemical treatment* is well established. The clinical waste is shredded, then exposed to the chemical agent (for instance sodium hypochlorite, chlorine dioxide, or peracetic acid). Some chemical systems use heat to shorten the treatment cycle.

- *Irradiation* can treat clinical waste, but needs extensive shielding to protect the users, and can only treat relatively small quantities of waste.

- In *encapsulation*, the waste container holds chemical packets whose contents, activated by the addition of liquid, encapsulate the waste in solid blocks.

CLINICAL WASTE INCINERATION

8.40   Modern clinical waste incinerators, properly managed, deal effectively with a wide range of clinical waste. Suitably designed and run municipal waste incinerators may also be authorised to deal with certain types of low risk waste, which is fed in sealed packages independently of the municipal waste stream.

8.41   In the past, most hospitals operated their own incinerators, not always to the highest standards. Hospital incinerators now have to meet new and demanding emission standards. The number of major incinerators treating clinical waste in England and Wales has gone down from approximately 700 to 37. The majority of these larger incinerators, which may be situated in hospitals, are now operated by private sector specialist companies.

LANDFILLING OF CLINICAL WASTE

8.42   The landfill directive will require all wastes to be treated prior to landfilling. All wastes that have the hazardous property *H9 Infectious* will be banned from landfill. The practical effect will be that any clinical waste must be adequately treated before landfilling can be considered as an option. Waste medicines, however, are classified as special waste: they must first be incinerated or otherwise treated by a process that denatures them. More generally, clinical waste classified as special waste is not suitable for landfill and – if no other pre-treatment is suitable – should be incinerated. Sanpro wastes, such as nappies and incontinence pads, may at present be landfilled without prior treatment.

Government and industry initiatives

A significant number of reports and guidance on the management and treatment of clinical waste have been published since 1995. These include:

- *Getting Sorted*, by the Audit Commission, which focused on segregation of hospital waste

- *Health Technical Memorandum 2075*, which gives advice on the selection, commissioning and operation of alternative clinical waste treatment technologies

- *Health Technical Memorandum 2065*, which focused on the segregation of waste stream in clinical areas

- Environment Agency's *guidance* on waste management licensing for clinical waste management – forthcoming

- *Safe Disposal of Clinical Waste*, published by the Health and Safety Executive, provides joint Environment Agency and HSE guidance on best practice for dealing with health care wastes.

Construction and demolition wastes

8.43 Construction and demolition waste represents a significant proportion of total waste generation. In 1989 it was estimated that 70 million tonnes of construction and demolition waste, including clay and sub-soil, arose annually. A study published in 1994 found that 29% of the 70 million tonnes was used on-site or sold off-site possibly after coarse crushing, 30% used for landfill engineering and 30% deposited at landfill. Only 4% was crushed to a graded product, the remaining 7% was disposed of illegally or used for agricultural purposes.

8.44 More accurate information on construction and demolition waste production, the quantities recycled or disposed to landfill, and the extent of variations between different areas in waste management practices, will become available after the completion of the survey being undertaken in March 2000 by the Environment Agency with the support of DETR and the National Assembly for Wales.

8.45 The need to reduce waste at all stages of construction was central to the message of *Rethinking Construction* – the 1998 report of the Construction Task Force on the scope for improving the quality and efficiency of UK construction. Improving the efficiency of the construction industry is a key objective for the Government, as set out in its strategy for more sustainable construction. The strategy, published in April 2000, identifies priority areas for action, and suggests indicators and targets to measure progress. It sets out action that the Government has already taken and further initiatives that are planned, highlighting what others can do. The Government will use the strategy as a framework to guide its policies towards construction, and will encourage people involved in construction to do the same.

8.46 The sustainable construction strategy emphasises the importance of reducing waste at all stages of construction by focusing on the need to consider long term impacts of design, construction and disposal decisions so that material and other resource use is optimised. The strategy encourages the industry (including its clients) to consider refurbishment or renovation as an alternative to new buildings and structures. It highlights the need to avoid over-specification in materials and the scope for standardisation of components.

8.47 The construction industry itself is rising to the challenge of achieving greater efficiency and sustainability through initiatives including the Movement for Innovation, Sustainable Construction Focus Group and the Construction Best Practice Programme. The sustainable construction strategy provides the framework for future action and achievement.

REDUCING AND RECYCLING CONSTRUCTION AND DEMOLITION WASTE

8.48 Some 260 million tonnes of minerals are extracted for use as aggregates and raw material for construction each year. Recycling construction and demolition waste as a substitute for primary aggregates has a double benefit – reducing both the amount of this waste which is landfilled, and the environmental impacts of quarrying primary minerals. Government policy on the use of construction and demolition waste as aggregates in England is currently set out Minerals Planning Guidance Note 6 (MPG6), published in April 1994. Aggregates and products made from aggregates should be recycled wherever possible and where technically, economically and environmentally acceptable, construction and demolition wastes should be used instead of primary materials.

8.49 MPG6 set targets for the use of construction and demolition wastes and secondary aggregate materials (such as colliery spoil, PFA, china clay and steel slag) for use as aggregate. The aim was to roughly double their use from about 30 million tonnes per year in 1989 to about 55 million tonnes per year by 2006, with a target of 40 million tonnes per year by 2001. Evidence from recent research suggests that there has been a substantial increase in the amount of construction and demolition waste used as aggregate. When the other secondary materials are included it is very likely that the MPG6 target for 2001 has been met. MPG6 is under review, and, subject to consultation, the revised note due during 2001 will help to improve the environmental performance of the aggregate minerals sector of the construction industry.

8.50 In Wales, minerals planning guidance is also being reviewed, and a consultation document was issued in November 1999. There are indications that there is less recycling of construction and demolition waste than in England, probably because of the availability and relatively low cost of primary aggregates. The consultation draft of the *Minerals Planning Guidance (Wales) Planning Policy* considers that one constraint on recycling has been the lack of facilities provided for the processing and storage of construction and demolition waste material. It proposes that development plans should indicate acceptable locations for recycling plants so that within 5 years every local planning authority should have made provision for recycling centres for construction and demolition waste. The draft MPG(Wales) Planning Policy is being revised following the end of the consultation period. It is intended to issue the policy guidance and the first two draft Technical Advice Notes – including one on aggregates – for consultation in 2000.

Welsh School of Architecture: construction waste management project

The project, which is part-funded by the European Regional Development Fund, targets construction waste management by providing counselling to building teams on the construction process, construction resources, waste reduction, and materials information exchanges. Besides the Welsh School of Architecture, the project team includes environmental consultants and industry representatives. A steering group has been set up to assist and advise the project team and includes representatives from industry, the Environment Agency, the National Assembly for Wales and local authorities.

The major objectives of the project are to:

- improve the life-time environmental impact and performance of buildings in the long term
- improve the environmental impact of construction operations through waste reduction at briefing, specification and site stages
- obtain data for benchmarking purposes

Project teams are being recruited and individual programmes of work drawn up for each participating company. Monitoring will take place throughout with assessments of activity taking place before, during and after each demonstration project. The lessons learnt will be widely disseminated to the construction industry through seminars, newsletters, websites, workshops and case study material.

8.51 Aggregates are not the only material wasted in construction and demolition processes. All materials are wasted to a degree. Furthermore, some construction and demolition wastes may be classed as special wastes. Measures to reduce waste in construction and to support the growth in recycling include:

- the strategy for sustainable construction

- the review of minerals planning guidance in England and in Wales, with supporting research – including work by consultants to identify ways and means of monitoring aggregate recycling

- introduction of a trial Aggregates Advisory Service that operated successfully from January 1997 to February 1999 with consideration being given to the arrangements for a longer term service

- introduction of an ongoing Internet based waste exchange to provide builders, civil engineers and demolition contractors with the opportunity to trade in surplus construction and demolition materials (Materials Information Exchange)

- forthcoming publication of good practice guidance on mitigating the environmental effects of recycling aggregates operations

- joint funding of a research project on the use of industrial by-products in road pavement foundations, now published

- joint funding of *The reclaimed and recycled construction materials handbook* published by the Construction Industry Research and Information Association (CIRIA) in 1999

- joint funding of research by CIRIA to publish guidance on waste reduction and recycling in construction – targeted as a site guide, design manual and boardroom handbook

- landfill tax on inert waste – charged at £2 per tonne of inert waste

- the introduction of an Aggregates Levy in 2002 (see section 8.103 of this part of the strategy for more information)

8.52   The waste, demolition and aggregates industries have responded positively to the policy of encouraging the use of recycled materials. The structure of the recycling industry is expanding and changing rapidly, spurred on by the tighter waste regulations and the landfill tax. This enthusiasm is being captured in developing the industry strategy for more sustainable construction.

Flood defence

The Environment Agency has dramatically increased the level of recycled aggregates used in flood defence works over recent years. The Environment Agency used over 800,000m³ of construction materials in 1998/9, of which 38% was secondary or recycled construction aggregates. For example, 500 tonnes of revetment stone from the Albert Dock flood defence scheme were reused to protect the Humber bank at Easington. 300 tonnes of recycled coping stones were used as sea defence to protect cliffs at Barmston Sea End from erosion. Brick rubble was used in place of new crushed stone on the Beck river works project in North East Region. Sand and ballast from a river channel alignment was used to build the sub-base for an 800m motorway diversion on the M4.

SPECIAL CONSTRUCTION AND DEMOLITION WASTES

8.53   A number of waste streams arising from construction and demolition works may be special waste e.g. contaminated soils, asbestos, tar and tar products, treated timber, paint and varnish. Wastes from the construction and demolition industry accounted for around 1.25 million tonnes in 1997/8 and 976,000 tonnes in 1998/9 – about 25% and 20% respectively of the total special waste consigned in England and Wales. This included some 500,000 tonnes of contaminated soils per annum and 356,000 tonnes (1997/8) and 162,000 tonnes (1998/9) of mixed construction industry waste (mostly contaminated with asbestos). Treated timber may also sometimes be classified as special waste.

8.54   Paint waste is generated by construction activities and some paints are special wastes. Careful ordering can reduce waste paint. Waste paint can be re-used either by the construction company for the next job or through community repaint schemes. Between 30,000 and 40,000 tonnes of waste paint was generated in 1997/8 and 1998/9, although it is not possible to say how much of this came from the construction and demolition sector. PVC is also a significant waste stream from demolition works (from windows and doors etc.).

ASBESTOS WASTES

8.55   All forms of asbestos can be carcinogenic, and all wastes containing greater than 0.1% asbestos are classified as special waste. Cement or bonded asbestos is a major hazardous waste stream. 242,000 tonnes and 190,000 tonnes of cement asbestos were consigned as special waste in England and Wales in 1997/8 and 1998/9 respectively. A further 146,000 tonnes and 102,000 tonnes of fibrous asbestos were consigned in England and Wales in 1997/8 and 1998/9 respectively.

8.56   All sources of asbestos wastes, including those from households, are classified as special waste. Re-use and recycling are not suitable options for asbestos wastes, as asbestos is banned for use in virtually all new applications. Treatment of asbestos wastes is difficult and expensive, although technologies to destroy asbestos using heat treatment or acid digestion are being explored. In the meantime the only practical option for asbestos wastes is disposal to landfill.

8.57 Asbestos wastes must not be allowed to enter construction and demolition waste streams destined for recycling as secondary aggregates. Crushing of asbestos wastes or their use in roadways or as hardcore will result in the wastes breaking up and could release fibres to the atmosphere. There is a shortage of facilities for the acceptance of asbestos cement wastes in some areas. Local authorities have a responsibility to provide sites for residents to deposit household wastes including certain asbestos wastes. Local waste development plans should consider the need for additional facilities to handle small amounts of commercial and industrial asbestos wastes. A lack of facilities may lead to an increase in fly tipping of these wastes.

Electrical and electronic equipment

8.58 The Government has encouraged the electrical and electronic equipment industry to take action to increase recovery and recycling rates for this sector. Some promising initiatives have been undertaken by industry. For example, there have been a number of trial collection schemes carried out or supported by organisations such as the Industry Council for Electronic Equipment Recycling (ICER) and the Electronic Manufacturers Equipment Recycling Group (EMERG). Under the umbrella of ECTEL (the European Trade Organisation for the Telecommunications and Professional Electronics Industry), a takeback project for mobile phones has been developed.

8.59 There have been no detailed analyses of the electrical and electronic equipment waste streams, so data on the amount of electrical and electronic equipment entering the waste stream has been estimated using predicted product lifetimes and market saturation information. Estimates indicate that around 1 million tonnes of this equipment is entering the UK waste stream each year. It is estimated that large household appliances, such as refrigerators, freezers and washing machines, and IT equipment comprise more than 80% of this amount. Consultants have predicted annual growth rates of 3-5% in the amount of waste produced. In consequence, although waste electrical and electronic equipment is estimated to be less than 1% of the total waste stream at present, this proportion may grow in future years.

8.60 Recovery and recycling rates vary, but data on these is relatively poor. A study carried out in 1992 estimated that over 75% of white goods are fragmentised to recover their ferrous and non-ferrous metal content, but recovery and recycling rates for most other types of equipment are likely to be significantly less than this. Collection trials throughout Europe have collected typically 30-40% of large items and 10% of small items, although there is a wide variation between the results of different trials. For some types of equipment, notably IT equipment, the markets for refurbished equipment are growing.

8.61 Collection, recovery and recycling infrastructures are being developed. Many retailers offer free takeback arrangements for large household appliances when the consumer buys a new appliance, and civic amenity sites also take old appliances. The experience of the pilot takeback project for mobile phones undertaken by ECTEL demonstrates that it is possible to recover smaller products, although there is no available takeback route for recovering and recycling most other small items of electrical and electronic equipment. In the commercial sector, old equipment may be taken back by the supplier as part of a contractual arrangement.

8.62 Scrap electrical and electronic equipment was chosen as an EC priority waste stream in 1991. The European Commission is developing proposals for an EC Directive on waste from electrical and electronic equipment. The proposed Directive would aim to reduce waste from electrical and electronic equipment, promote the re-use, recovery and recycling of this equipment, and reduce the impact on the environment of treating and disposing of this equipment.

8.63 The European Commission's draft proposals include quantified targets for collection, re-use and recycling of electrical and electronic waste, and a proposal that equipment manufacturers and importers should bear the costs of collecting, treating, recovering and disposing of household equipment. The draft proposals also include measures aimed at reducing the use of hazardous substances in electrical and electronic equipment. The scope of the proposals is wide, covering large household appliances, small household appliances, IT equipment, telecommunication, radio, television and electro-acoustic equipment, musical instruments, lighting equipment, medical equipment systems, monitoring and control instruments, toys, electrical and electronic tools, and automatic dispensers.

8.64 The Government supports the broad objectives of the Commission's proposals, and will participate constructively in negotiations on the Directive. The Government has commissioned work which is helping to assess the likely environmental benefits and financial costs that would arise from the proposals being put forward by the European Commission. The outcome of this assessment is informing the Government's views on the Directive.

8.65 A number of electrical and electronic equipment waste streams were added to the EC Hazardous Waste List in December 1999, and will need to be treated as special waste in the UK from no later than January 2002. These include equipment containing chlorofluorocarbons as well as equipment containing PCBs, free asbestos and other hazardous components.

8.66 Many electrical and electronic goods contain hazardous components even if they themselves are not classified as hazardous waste. Figures from Denmark suggest waste electrical and electronic equipment may account for as much as 60% of copper and 20-40% of lead going for disposal. TV screens and monitors contain cathode ray tubes, which in turn contain a number of hazardous components including cadmium, phosphor and arsenic. The landfill of untreated cathode ray tubes may be forbidden under the Waste Electrical and Electronic Equipment Directive, and companies are developing methods of recovering materials from cathode ray tubes that allow the hazardous substances to be removed and the various materials (plastic, metal and glass) to be separated.

End-of-life vehicles

8.67 Each year over 1.5 million vehicles reach the end of their lives in the UK, either because of their age (typically around 12 years for cars) or because they are heavily damaged in an accident. An average of about 75% of the weight of each end-of-life vehicle is presently recycled, mostly through the re-use of parts and metals for recycling. Recycling other materials, such as plastic, glass and rubber is more difficult for technical and economic reasons and currently these end up mostly as landfill and account for about 0.3% of total UK controlled waste production.

8.68   The Automotive Consortium on Recycling and Disposal (ACORD) was set up in 1991 and includes vehicle and material manufacturers, dismantlers and material recyclers. In 1997, ACORD members signed a voluntary agreement committing themselves to making significant improvements in this area and in particular setting targets to improve the recovery of material to 85% by weight by 2002 and 95% by 2015.

8.69   To achieve these goals, vehicle manufacturers agreed to work on making their vehicles more suitable for recycling; dismantlers agreed to develop improved dismantling and draining processes; metal recyclers agreed to promote the reduction of end-of-life vehicle wastes; and the plastics and rubber industries agreed to develop material recycling technologies and market materials and products with recycled grades.

8.70   ACORD is also working closely with another motor industry initiative (known as CARE) which, for the last five years, has been working on a number of technical pilot projects aiming to improve recycling technology and develop markets for recycled materials.

8.71   The improvements expected from the ACORD and CARE initiatives will be vital if industry is to meet the requirements of the proposed EC Directive on end-of-life vehicles. This directive differs in certain areas from the ACORD agreement but has the same central objective of reducing the amount of waste generated from end-of-life vehicles.

8.72   The End-of-Life Vehicles Directive has its origins in the work of a Priority Waste Stream Working Group, set up by the Commission in 1991, which looked at ways of reducing the amount of waste from end-of-life vehicles. It aims to meet its environmental objectives by:

- setting reuse, recycling and recovery targets. It would increase the reuse and recovery rates to 85% by 2006 and 95% by 2015

- requiring manufacturers to design and manufacture their vehicles with recyclability and reuse in mind, and to restrict the use of heavy metals such as lead

- requiring Member States to set up systems to ensure that all end-of-life vehicles are captured by an approved treatment chain. This would include introducing certificates of destruction issued when the vehicle is discarded by the last owner

- setting treatment standards which authorised dismantlers and other facilities must meet

8.73   The End-of-Life Vehicles Directive is presently in draft form and although the Council and The European Parliament have not yet agreed on the final form of the Directive, we expect a decision to be reached during 2000.

8.74   The End-of-Life Vehicles Directive is likely to require that vehicles are drained and that hazardous materials or components are removed before treatment. These include petrol, diesel, engine oil, automotive batteries and brake fluids. Around 150,000 tonnes of engine oil are treated as special waste annually, alongside 140,000 tonnes of automotive batteries. Currently only limited quantities of other materials are treated as special waste – although this may increase as more separation of the waste into its constituent materials takes place, to meet the Directive's requirements.

8.75 The Environment Agency has produced guidance on the classification of materials and liquids from cars. Many of the components from vehicles are suitable for recycling and recovery. Oils can be regenerated or, if this is not practical, they can be used as a fuel. More than 90% of automotive batteries are recycled (see Chapter 8 sections 8.18 of this part of the strategy). Oil filters can be sent for recycling of the metal content and recovery of the energy from the oil. In the Netherlands the landfill of oil filters has been banned to encourage recycling. The vehicle shells can be recovered usually following shredding and sorting, as a source of secondary metals.

Glass

8.76 Soda-lime glass is produced by heating a mixture of sand, soda ash and limestone in a furnace to a temperature in excess of 1500°C. The raw materials used in the production of glass are relatively cheap and plentiful, but large amounts of energy are needed to convert them into glass. The manufacture of glass containers, such as bottles and jars is the largest use of glass and the manufacture of flat glass, such as windows and car windows, is the second largest use. Other types of glass, such as borosilicate glass, have a different composition to soda-lime glass, but only make up a small percentage of total production.

UK consumption and recycling of glass container cullet				
Year	Estimated total consumption (tonnes)	Estimated total recycled (tonnes)	Percentage recycled	No of sites in UK Bottle Bank Scheme
1993	1,810,000	395,000	21.8%	10,965
1994	1,800,000	404,000	22.4%	12,858
1995	1,900,000	412,000	21.7%	14,300
1996	2,000,000	430,000	21.5%	15,609
1997	2,100,000	440,000	21.0%	19,341
1998	2,200,000	476,000	21.6%	–

8.77 Reusable bottles are designed to be repeatedly returned to the manufacturer, cleaned and then refilled. These bottles have to be stronger and heavier to allow for multi-use. This increases both production costs, transport costs and energy use. Their re-use relies on the efficiency of the collection and return system.

8.78 Recycling involves the collection of waste glass bottles and jars, crushing them into cullet which is melted with virgin material in a glass-manufacturing furnace. The recycled glass is separated by colour, usually at the collection point, and all non-glass materials are removed prior to crushing.

8.79 The main barrier to the recycling of glass is the shortage of brown and white cullet in the UK. There is still a large imbalance between the UK's predominantly white flint (clear) and amber glass production and the high proportion of waste green glass, which consists of mainly imported wine and beer bottles.

8.80 The percentage of glass in an average household dustbin is estimated to be around 9%. Current legislation should help to reduce this amount:

- the landfill tax, which is charged on a tonnage basis, is expected to encourage local authorities to develop alternative re-use and recycling schemes for glass wastes

- the packaging Regulations means that the UK will need to improve its glass recycling performance, with over 300,000 tonnes of packaging and container glass being recycled by 2001

The Bottleback Initiative

The alcoholic drinks industry has set up a new initiative, called Bottleback, which is aimed at recovering and recycling non-returnable bottles (NRBs) from licensed premises, such as clubs, pubs and restaurants. There has been a considerable rise in the number of non-returnable bottles sold in the UK over the last 10-15 years due largely to the fashion for drinking from the bottle.

The Brewers and Licensed Retailers Association (BLRA) estimate that there is 350,000 tonnes of this glass outside licensed premises. Most of this glass is not separated from other trade waste and consequently goes to landfill. Some green glass will be collected, but the colour of most of the glass is more likely to be clear and amber/brown. It is these colours which are needed by the glass industry, as they are in short supply compared to green glass.

A large number of licensed premises are involved in the service. The principal operators, Biffa and Cleanaway, have projected an optimum route density for their vehicles to ensure collection occurs at the lowest possible cost. All licensed premises on the optimum route are encouraged to join the scheme. The alternative to this would lead to vehicles travelling large distances to collect from the diverse premises of a specific pub estate owner.

The advantages for licensed premises owners are a reduction in the cost of waste disposal and compliance with the packaging Regulation requirements. Bottleback is currently operating in London, Manchester, Birmingham and South Yorkshire and is expected to expand across the country during 2000.

Green waste

8.81 Green waste includes vegetation and plant matter from household gardens, local authority parks and gardens and commercial landscaped gardens. Kitchen wastes refers to food preparation waste and plate scrapings. The annual production of green waste is estimated to be around 5 million tonnes per year.

8.82 Composting of green waste is a viable alternative to landfilling. In 1997, there were a total of 39 fully operational composting sites in the UK, with a further 12 being partly operational, under construction or planned. In addition to composting green waste, some of these sites also accept source separated kitchen waste.

8.83 Composting of green waste is not as widely used as it might be for two reasons in particular:

- the low cost of landfill relative to composting: until recently, it has not been cost effective for waste management companies to select composting over landfill

- the lack of a market for the product: the principle barrier to more widespread use is the negative perception of composted waste as a product, and the problem of the absence of accepted standards for waste derived compost

8.84 The year on year increase in the landfill tax, and the Landfill Directive, are expected to have significant positive impacts on composting as a waste management option:

- the impact of the landfill tax will increase the economic viability of composting when compared to landfilling. This should promote its development by waste management companies and its adoption by local authorities

- the Landfill Directive specifically targets the biodegradable fraction of municipal waste. An increase in the extent of composting of green waste will probably be required if the targets set out in the Directive are to be met

8.85 Obtaining a consistent quality product from composting requires a consistent quality waste feedstock with little or no contamination. Composting plants have frequently been beset with problems over odour and public nuisance, particularly where controls on inputs and/or operational practices are inadequate. In order to ensure composting is a success it needs to be carefully managed, and can require expensive capital investment for shredders, plant, drainage, and roofing and enclosed housing. Further details on composting can be found in Chapter 5 section 5.32 of this part of the strategy. Alternatives for green waste include incineration with energy recovery, and chipping or shredding to create mulches.

Metals

8.86 Scrap metal is derived from two main sources:

- *new scrap* is scrap derived from metal processing, such as off-cuts, stampings, turnings, grindings and swarf from industries carrying out metal fabrication processes. Almost 100% of this waste metal is recycled

- *old scrap* is metal derived from end of life or obsolete products. This includes *heavy scrap* from dismantling industrial plant, railway rolling stock and track, and also *light scrap* from the processing of consumer goods. Scrap metal from end of life vehicles is discussed further in section 8.67 of this Chapter

8.87 Recycling is often the Best Practicable Environmental Option for waste metal, and scrap metals represent by far the largest volume of industrial material that is recycled. Recycling provides a high grade feedstock to refining processes which reduces the use of raw materials, energy and the quantity of residues arising from the process. The costs of the recovery, collection, sorting by metal type, and removal of unwanted impurities from waste metal tends to be favourable to the higher costs of producing metal from metal ore.

Scrap re-used as a percentage of consumption

(Note that re-use is not the same as recycling – these figures represent the amounts of metal sourced from existing products, not the amount of each metal recycled each year)

Year	Iron	Lead	Copper	Zinc	Aluminium
1992	45%	64%	35%	21%	39%
1993	42%	67%	35%	21%	29%
1994	42%	74%	32%	20%	39%
1995	40%	71%	34%	20%	53%
1996	44%	73%	36%	19%	44%
1997	45%	69%	37%	19%	40%
1998	39%*	66%	38%	18%	43%

*This drop to 39% compared to the previous yer is due to a surge of steel industry products during 1998

8.88 Scrap metal wastes are mostly collected through a well-established infrastructure, passing from the smaller scrap metal yards to the main dealers. At each stage in the chain, the scrap is sorted to remove high value non-ferrous items and bulked into standard classes of material. Large items are broken down using processes such as cutting, compacting and fragmentising, each producing a particular grade of scrap metal.

8.89 Metal can recycling has become common practice, but it is a complex operation. Steel cans are tinned and lacquered internally to protect the contents. Some steel drinks cans have aluminium ring pulls and beer cans now contain plastic widgets. All have an external coating of coloured markings and residues of the contents. All used can processing involves shredding, material separation and decontamination stages.

8.90 The recycling of aluminium cans has increased significantly in the UK, from 16% in 1992 to 36% in 1998. This is mainly due to the aluminium industry having established collection schemes and reprocessing facilities for their used products. Even so, collection of waste aluminium cans is currently insufficient to fully utilise the available reprocessing capacity in the UK, but the recovery and recycling targets in the packaging Regulations are likely to result in further increases in collection rates.

8.91 Iron can be recovered from the ash of municipal waste incinerators. This iron is not as valuable to the steel industry as that recovered via can-banks. It contains residual tin levels (up to 0.25%) which cannot be further reduced because the high temperature incineration alloys the tin with the iron. The amount of iron recovered via this route has decreased since 1994, due in part to the closure of some incinerators following the introduction of the Waste Incineration Directives requiring lower emission levels.

Friction stir welding

The Welding Institute has developed a new technique called *friction stir welding* for welding aluminium alloy. This alloy – a mixture of aluminium and other metals – is likely to be a major material of the future, being light in weight, strong and easy to manufacture. It is also theoretically possible to recycle the material at around 5% of primary production cost. However, current welding techniques for aluminium alloy are costly, energy intensive and potentially harmful to the environment.

The friction stir welding process uses a special tool to heat the aluminium only as far as its plastic heat, making it flexible. The process does not require a shielding gas or filler wires. It is claimed that the need for fume extraction and filtering equipment is unnecessary for the process, which also does not produce any radiation hazards. The process should increase the scope of aluminium alloy application in the future. Furthermore, the process apparently only consumes 2.5% of the energy needed for a laser weld – representing a significant saving in CO_2 emissions from power generation and an additional cost saving for industry.

8.92   The packaging Regulations means that the UK has to improve its metal recycling performance by 2001. Both steel and aluminium must meet 15% material specific recycling targets within the 50% overall packaging waste recovery target – some 110,000 tonnes of steel and some 16,000 tonnes of aluminium in 2001.

METALS AS SPECIAL WASTE

8.93   Most metals are not hazardous in their metallic form, although there are exceptions such as mercury and thallium. However, many metals form hazardous compounds which may be present in scrap, for example lead plates from automotive batteries contain significant quantities of lead oxide and lead sulphate, both of which are hazardous.

8.94   Recycling for some heavy metals can be difficult. For example, there is a limited market for recycled cadmium, as cadmium is a by-product of zinc manufacture. The use of cadmium in products such as paints, pigments and batteries is declining as other more environmentally acceptable technologies are developed. Recycling of cadmium-bearing wastes is therefore uneconomic unless they contain other valuable raw materials. Similarly the use of mercury is being phased out in many applications, thus leaving little incentive for recycling mercury as the markets for the product are shrinking. There are strong environmental reasons for segregating mercury from waste streams, but its long-term management could become a problem. Mercury compounds are generally more toxic than mercury metal, and therefore long term storage as metallic mercury may be the Best Practicable Environmental Option at this time.

Mine and quarry wastes

8.95   Mine and quarry wastes include materials such as overburden, rock inter-bedded with the mineral, and residues left over from initial processing of the extracted material into saleable products. Extraction and processing waste may include:

- materials such as waste rock and sandy debris
- fine grained materials (tailings), derived from crushing and washing the mineral

8.96   These materials are non-hazardous and mostly chemically inert. They are often largely identical to the geological deposits in the locality from which they are extracted.

8.97   Mine and quarry wastes have been produced, in particular, in connection with the extraction of coal, clay, gravel, chalk, slate and other metals and minerals with commercial value. Where wastes would impede mineral extraction if returned to the mine or quarry, large tips may accumulate. While wastes can be left underground in some types of mine, this is not always economically feasible hence the large surface tips associated with underground mining of coal. The return of china clay waste to the pit is limited by the need to keep access to a variety of grades of clay for blending purposes.

8.98   Tips may take up significant areas of land as well as being large and unsightly. A 1994 survey of land for mineral workings in England indicated that just under 14,500 hectares of land had planning permission for surface disposal of mineral working deposits, nearly 9,900 were actually affected by tipping (although some sites were in the process of being reclaimed), and that over 6,000 hectares had planning conditions for reclamation at that time.

8.99   Many of the larger surface tips arose in the past and while some are being added to, others were abandoned before modern controls on mineral working came into force. Some old tips may be re-worked for minerals (such as the extraction of vein minerals from tips in Derbyshire and the extraction of coal from some old colliery spoil tips), which can often result in the rehabilitation of such areas. Land reclamation under derelict land grant and, more recently, through the reclamation programmes of English Partnerships and the Welsh Development Agency, has dealt with many old colliery spoil tips and some other mine tips which had no requirements for reclamation or restoration.

8.100   Much of the waste arising at surface mineral workings is used for infilling prior to restoration of the land to a beneficial subsequent use. This is particularly the case where large amounts of overburden and interbedded materials have to be moved to get to the mineral, for instance in opencast coal operations, or extraction of fuller's earth or ball clay. Such restoration is a key element of sustainable working of those minerals, and is a requirement of the planning permission for extraction.

8.101   The amounts of mine and quarry wastes produced annually are not known accurately but, in the early 1990s, stood at about 110 million tonnes per year. The annual amounts arising are likely to have fallen since then, in particular because of a decrease in coal extraction. It is thought that about 5% of annual production of mine and quarry wastes are re-used, mainly in the construction industry. Because the amount of waste generated is relatively large, and re-use is limited, considerable amounts of material have accumulated over the years. The resulting tips can be environmentally intrusive although, in some cases, natural regeneration or restoration has turned some of these into conservation assets.

8.102   Many of these materials do have a potential for greater use as construction fill or, sometimes, as higher quality aggregate. More utilisation of these, where standards and specifications permit, would help to partly offset the need for extraction of aggregate minerals from the ground and would help to reduce the extent of land required for tipping. Whilst recognising that tipping and restoration will continue to be the main management option for mine and quarry wastes for the foreseeable future, the Government and the National Assembly wish to maximise the re-use of such materials where it would not conflict with either:

   •   environmentally beneficial restoration of the mineral working

   •   naturally regenerated sites which have amenity or conservation value, or sites where the removal of material from the site might have other significant environmental impacts

8.103   There are environmental impacts associated with aggregate extraction. The quarrying industry can influence these impacts through the methods it uses to extract and transport materials, while others further down the construction chain can help to reduce the demand for primary materials by good design and specification of materials, by tighter control of waste on sites, and by re-using and recycling wherever possible. The Government has said that an Aggregates Levy will be introduced in 2002 to reflect the environmental costs of aggregates quarrying and encourage demand for, and supply of, alternative materials such as mineral wastes and recycled construction and demolition waste. The Levy will also help tackle the present high levels of waste in the use of construction materials.

LAND USE PLANNING

8.104   The use of land for mineral extraction and the control of minerals development is regulated under the Town and Country Planning Act 1990, as amended by the Planning and Compensation Act 1991. Minerals Planning Authorities are required to prepare development plans for minerals which set out the policies and proposals against which planning applications are determined. Consideration includes all aspects of mineral working including provisions for the placing of mineral wastes in tips and lagoons, and restoration proposals. Guidance on minerals waste is contained in *Minerals Planning Guidance Note 1 General Considerations and the Development Plan System*. This encourages the use of minerals wastes where this is practicable and justified, and the inclusion in development plans of policies for recycling as well as identifying areas for future disposal of mineral wastes.

8.105   In Wales, minerals planning guidance is being reviewed. Draft *Mineral Planning Guidance (Wales) Planning Policy* was issued for consultation in November 1999. One of the key areas in the draft guidance relates to encouraging the efficient use of minerals and reducing the production of waste by maximising the potential for re-use and recycling of materials where environmentally acceptable. The draft policy guidance is being revised following the end of the consultation period, and it will be issued with two of the draft Technical Advice Notes for consultation in 2000.

8.106   In the case of tipping at existing mines under the Town and Country Planning (General Permitted Development) Order 1995, limitations are placed on the size by which a tip can be increased unless otherwise provided for in a waste management scheme which, among other things, would provide for the restoration and aftercare of the site. Under the Order, the tipping of colliery spoil on waste sites already in use for that purpose in 1948 can continue, provided it is in accordance with a scheme approved by the mineral planning authority. All other mineral waste tipping requires a new planning permission.

8.107   The 1995 Environment Act provided for an initial review and updating of all old permissions for development consisting of the winning and working of minerals or involving the depositing of mineral waste, and the periodic review of such permissions thereafter. The aim is to ensure that planning conditions reflect changing environmental standards and do not become outdated. Advice was issued in *Minerals Planning Guidance Note 14 Environment Act, 1995: Review of Minerals Planning Permissions*. Together, these provisions ensure that all tipping operations are subject to proper planning control.

ROLE OF LOCAL AUTHORITIES

8.108   The Minerals Planning Authority is responsible for strategic planning for the provision of minerals and for deciding applications for extraction of minerals, including provisions for tipping of minerals wastes and for rehabilitation of sites. The planning authority will set planning conditions to control all relevant aspects of the operation.

8.109   The proposals for dealing with wastes may be a material consideration in deciding an application and it may be appropriate to consider to what extent wastes might be recycled, rather than tipped, and whether re-use of these materials might inhibit proper restoration of the site. In considering policy for provisions of aggregates, Minerals Planning Authorities should also consider whether mine and quarry waste can make a significant contribution to meeting demands for construction fill and aggregates, and the implications of this for permitted reserves of primary aggregates.

Waste oils

8.110   This section covers waste mineral oil from automotive, industrial and other sources. Most lubricating oil contains additives which produce a specified performance from the oil. Additives may include rust inhibitors, detergents or alkaline compounds, and constitute between 5% and 25% of their formulation. Waste oils can contain traces of the additives and contaminants following use, including metals or combustion products.

8.111   All waste mineral oils are classified as special waste, although small quantities (5 litres for disposal, 20 litres for recovery) may be moved without consignment notes, provided that there is a minimised risk of pollution to the environment or harm to human health. Improved refining and formulation have reduced the hazardousness of many oil products that are not classified as dangerous substances for supply. Oil is a highly polluting substance and was responsible for more pollution incidents than any other group of substances, according to the Environment Agency's pollution incident database for 1998.

8.112   There are a number of sources of waste oil, including: automotive engine, transmission and gear oils; industrial gear, hydraulic, compressor, transformer and cutting oils; and others including marine and aviation oils. The total quantity of lubricating oils sold in 1997 was approximately 872,400 cubic metres.

Estimated quantities of oil sold, and waste management option used			
	1995 (m³)	1996 (m³)	1997 (m³)
Total oil sold	895,000	864,329	872,378
Waste oil regenerated	32,000	32,000	32,000
Waste oil combusted after processing	390,000	390,000	390,000
Total Oil recovered	422,000	422,000	422,000
Landfilled (including permanent storage)	70,250	49,550	54,637
Unaccounted for/lost in use*	402,750	392,779	395,741

*A significant proportion of the oil that is unaccounted for is lost by combustion in engines or evaporation, therefore not all of this quantity is disposed of incorrectly

8.113   There are three main waste management options for dealing with waste oils: regeneration; combustion after treatment; and combustion without treatment. Article 3(1) of the Waste Oil Directive (75/439/EEC, as amended) requires priority to be given to the regeneration of waste oil where technical, economic and organisational constraints allow.

8.114   *Regenerating waste oil:* waste industrial lubricants can be regenerated through laundering, reclamation or re-refining:

- laundering is most appropriate for waste oils of a known composition. Laundering involves heating, filtration, de-watering and the addition of fresh additives before the oil can be re-used

- reclaimed waste oils can be used for secondary purposes, for example, as a mould release in foundries. Treatment may involve centrifuging and/or filtering to remove impurities

- waste oil can be re-refined into a base stock oil ready for blending. A number of processes are used, with varying success both in terms of environmental and economic performance

8.115   *Combustion after treatment:* waste oil can be combusted after treatment. Different levels of treatment can be applied to the waste oil depending on the desired application. Low level treatment allows waste oil to be blended into fuel oil which, for example, can be used in road-stone plants to dry limestone for road construction. This use accounts for 60 – 80% of combusted waste oil. Waste oil can also be treated to produce a fuel with similar properties and emissions levels to that of virgin fuel. Waste oil is passed through a flash column to remove water, and then, through distillation, sediments, heavy hydrocarbons, metals and additives are removed.

8.116   *Direct combustion of waste oils:* waste oil is also burned in small space heaters without any pre-treatment.

Government and industry initiatives

The Environment Agency's Oil Care Campaign aims to improve the way in which waste oil is managed by providing information on good practice. The Agency also run an Oil Bank Line which provides details of the closest oil recycling bank.

The campaign has been successful in raising the profile of oil collection facilities and at the same time the number of collection points has risen.

An initiative by the National Household Hazardous Waste Forum aims to double the number of oil recycling banks in the UK. One of the main objectives is to recover much of the unaccounted for waste oil that is produced by DIY mechanics.

Industry has also seen a gradual increase in the amount of oil recovered. This can be accounted for by an increase in the awareness of the environmental impact of oil and the spread of more effective environmental management systems.

8.117   The proposed Waste Incineration Directive is likely to apply stringent emission standards to any plant burning a wide variety of wastes, including waste oil. This will significantly restrict the number of existing furnaces where waste oil can be burned and, as a consequence, will probably reduce the amount of waste oil managed in this way.

8.118   The efficiency of collection of waste oils has improved over recent years, helping to reduce the amount of waste oil lost to the environment. As public and commercial awareness grows, the amount collected is set to continue to increase. A number of initiatives are helping to increase awareness.

Ozone depleting substances

8.119   Ozone depleting substances (ODSs) are a family of man-made chemicals used in a variety of applications such as refrigeration, fire fighting, foam blowing, and as solvents and aerosol propellants. Their success in these applications has been mainly due to their stability. However, this stability also allows these substances to escape the lower atmosphere into the stratosphere, where they may interfere with the natural life cycle of the ozone layer and cause its depletion. The most common of the ODSs are chlorofluorocarbons (CFCs), hydrochlorofluorocarbons (HCFCs), halons, carbon tetrachloride, 1,1,1-trichloroethane, methyl bromide, and hydrobromoflurocarbons (HBFCs).

8.120   Because of concern about the future of the ozone layer, in 1987 the world community agreed the *Montreal Protocol on Substances that Deplete the Ozone layer*. To date over 170 counties, including the UK, have signed up to it. Within the EU the Protocol is implemented by EC Regulation 3093/94, which not only enforces the Protocol but also introduces even tighter controls. As a result of this Regulation, production and consumption of the most harmful ODSs were phased out between 1994 and 1996, and phase out schedules are in place for HCFCs. Recent revisions to the Protocol, however, mean Regulation 3093/94 will soon be replaced by a new EC Regulation. Key elements of the proposed new EC Regulation having implications on waste management policies would include:

●   an immediate ban on the sale and use of most CFCs, halons, carbon tetrachloride, 1,1,1-trichloroethane, HBFCs and bromochloromethane (CBM)

●   a further ban on the supply and use of CFCs used in maintenance or servicing of refrigeration and air-conditioning equipment from 31 December 2000, and on halon use in existing non-critical fire protection systems from 31 December 2002 (these halon systems would need to be decommissioned by the end of 2003)

●   compulsory recovery and disposal of ODSs in refrigeration and air conditioning, solvent and fire protection equipment (this would apply to other products, installations, and equipment where practicable, and domestic refrigeration equipment after 31 December 2001)

8.121   Recovery of these ODSs usually takes place during the decommissioning process and under the terms of the proposed new EC Regulation Member States will be required to define the minimum qualifications required of the personnel involved. As a system is replaced or retrofitted the supplier will generally remove the ODSs, taking it for recycling or destruction.

8.122   Ozone depleting substances that cannot be recycled must be destroyed. There are currently two ozone depleting substance disposal plants in the UK, both of which employ high temperature incineration. Because of the measures described above, there are likely to be relatively large quantities of these substances to be disposed of for several years after 2000, as refrigeration and air conditioning systems as well as fire protection equipment are decommissioned.

Government and industry initiatives

DTI and DETR have produced a number of guidance documents which aim to assist users in selecting alternatives to ozone depleting substances. These provide guidance on alternatives for CFCs, Halons and ozone depleting solvents and include:

- Refrigeration and Air Conditioning – CFC Phase Out: Advice on Alternatives and Guidelines for Users

- Fire Fighting – Halon Phase Out: Advice on Alternatives and Guidelines for Users

- Ozone Depleting Solvents – CFC Phase Out: Advice on Alternatives and Guidelines for Users

- Refrigeration and Air Conditioning: HCFC Controls

- Do you use HCFCs?

- Proposed new controls on HCFCs, 1,1,1 Trichloroethane, and Carbon Tetrachloride: UK Compliance Cost and Impacts Study

- New Controls on Ozone Depleting Substances

Updates to most of the above will be issued once the proposed new EC Regulation enters into force.

Paper

8.123   Waste paper, which includes boards, boxes and cartons, can be categorised into four main groups, according to their recovery use:

- *primary pulp substitute grades:* recovered papers in this smallest group can be used with very little processing to make new paper. They consist of mainly unprinted trimmings from printers and are used primarily to make printing and writing paper

- *de-inking grades:* these are grades from which the ink is removed before the recycling process begins. Newspapers and magazines are recycled in large quantities mainly for newsprint, while office waste is a key source of high quality grades used to make mainly new hygienic (tissue) and graphic papers

- *kraft grades:* the largest single group generally comes from brown unbleached packaging materials, such as corrugated cases. Their strong fibres make them essential for recycling to become new packaging

- *low grades:* consists of mixed papers which are uneconomical to sort but are usable for packaging and speciality boards

Main types of paper produced (UK, 1998)		
Type of Paper	**UK Production (tonnes)**	**Percentage of UK Production**
Newsprint	1,043,000	16.0%
Printings and writings	1,767,000	27.0%
Corrugated case materials	1,760,000	27.0%
Packaging papers	160,000	2.5%
Packaging board	667,000	10.5%
Household and sanitary	635,000	10.0%
Other paper	441,000	7.0%
Total	6,473,000	100.0%
Data for the UK (1997) provided by the Pulp and Paper Information Centre		

8.124   The preferred waste management option for waste paper is recycling, though energy recovery through incineration and landfill are also widely used.

8.125   Recycling of fibre is generally considered to be the priority and Best Practicable Environmental Option. Most recycled paper comes as used packaging from retailers and others (1.6 million tonnes) and once-read newspapers and magazines (1.25 million tonnes), while pulp substitutes and de-inking grades come from printers and offices respectively. These routes offer the most economical way of collecting the waste paper. The recovery rate has increased over the past 5 years, but is constrained:

- in 1998 the UK imported 60% of its annual paper requirement. This was 7.4 million tonnes of paper from overseas, with major (virgin) paper suppliers being Finland and Sweden

- over 4.6 million tonnes of all papers were recycled back into the UK's production of 6.5 million tonnes in 1998. Additionally, 0.4 million tonnes were exported, while imports were minimal

- it is not possible to recover all paper, for example 0.9 million tonnes is used for hygiene purposes, while other types are in effect *stored* as books, wallpaper and office files

- fibre strength decreases during recycling and individual fibes cannot be used indefinitely. The addition of virgin papers helps to maintain the overall quality of the recycled product

- waste paper is a world traded commodity and is subject to fluctuations in demand and price, as is virgin pulp. The latter plays a significant part in recovered paper prices

8.126   A recent independent study shows recycling of newspapers and magazines to be the preferred choice above incineration and landfilling – while fibres are strong enough to re-use. When the fibres become too weak to re-use, then the paper can be incinerated, both to maximise energy recovery and reduce landfill volumes needed. For either re-use or incineration, UK transportation costs are limited in terms of both economic and environmental impact.

8.127 The packaging Regulations require the UK to improve its paper packaging recovery and recycling performance, by 2001. Packaging paper waste volume in the UK in 1998 has been estimated as almost 4 million tonnes.

Polychlorinated biphenyls

8.128 The electrical and heat transfer properties of polychlorinated biphenyl (PCB) led to its widespread use by industry and in electrical products for commercial and domestic use (for example in capacitors and transformers). PCB is very resistant to chemical and biological degradation and has become widely dispersed, but at low levels, in the environment. It is also soluble in fats and oils and tends to accumulate in the fatty tissues of living organisms. It has been found in the body fats of predatory sea birds and sea mammals. PCBs have been linked with birth defects in animals as well as animal deaths through immune deficiency.

8.129 The use of PCBs and their equivalents has been progressively restricted since the 1970s. However, it is recognised that PCBs, which remain in existing equipment, pose a continuing environmental threat. The UK and other North Sea states agreed in 1990 to phase out and destroy remaining identifiable PCBs by the end of 1999. In 1996, the EU adopted the PCB Disposal Directive (96/59/EC). This requires all Member States to take measures to phase out and destroy identifiable PCBs, subject to certain derogations, by the end of 2010.

8.130 In 1997, the Government issued the UK Action Plan for the phasing out and destruction of PCBs and equivalents. The Plan was intended to advise industry and others with PCB holdings of the UK commitment made at the North Sea Conference in 1990 as well as the requirements of EC Directive 96/59/EC. The Action Plan also foreshadowed the making of Regulations to give legislative effect to these international commitments. The Government consulted on draft regulations in 1999. It is intended that the Regulations will be made before the end of March 2000.

8.131 The main waste management methods available for destruction of PCB's include:

- **high temperature incineration:** combustion in an incinerator running at a temperature of at least 1100°C

- **chemical dechlorination:** these methods of treating PCB-contaminated oils are generally based on reactions with sodium and potassium to form inorganic chlorine salts which can be removed by filter or centrifuge. The treated oil may be re-used in electrical equipment

- **hydrodechlorination:** treatment of PCB with high-pressure hydrogen gas in the presence of a catalyst to produce dechlorinated PCB and hydrochloric acid

8.132 Other techniques for separation or destruction of PCBs have been developed, including solvent extraction, wet air oxidation and electrochemical techniques.

8.133 Various alternative dielectric fluids based on environmentally benign substances are now available to replace PCB in electrical equipment, and modern transformer plant has been developed for some applications, which avoids the use of liquid dielectrics altogether.

National Grid Ltd

National Grid is a major company taking a responsible approach to the management of PCBs in their equipment. The company operates a mobile processing unit, known as PCB Gone, to destroy the PCB content of transformer oil. The unit uses chemical dechlorination, and produces PCB-free oil, which can be re-used. The plant treated 100,000 litres of oil in the year 1997/98. This brings the total processed in the three year period to 1998 to 1.7 million litres. The company has now set itself targets to decontaminate the oil in all its transformers containing PCB levels greater than 50 mg/kg by 31 December 1999.

Plastics

8.134   It is estimated that there are 2.8 million tonnes of various types of plastic waste generated in the UK annually. Packaging currently accounts for approximately 60% of this waste, although waste production in the construction and electronics sectors are expected to increase rapidly. Plastics account for approximately one fifth of waste packaging materials; the split between domestic and industrial plastic packaging waste is assessed at 1.2m tonnes domestic and 0.5m tonnes industrial. Plastics contribute about 11% of material in the household waste stream, although the perception can be considerably higher.

Percentage quantities of plastic types in household waste in Europe (AMPE)	
Low Density Polyethylene (LDPE)	23%
High Density Polyethylene (HDPE)	17%
Polypropylene (PP)	19%
Polystyrene (PS)/Expanded Polystyrene (EPS)	12%
Polyvinyl Chloride (PVC)	11%
Polyethylene Terapthalate (PET)	8%
Other plastic types	10%

8.135   The main disposal route of waste plastics is currently landfill, with some incineration and recycling. Traditionally, plastics have tended not to be recycled as the cost of recycling and decontamination have militated against the process. Recycling methods include:

- ***direct recycling to virgin plastic*** which is difficult due to contamination, mixture of grades and the type of original product, thus selective collection and sorting is required

- ***re-melting and extrusion***, using certain types of plastics as raw materials for making objects such as vehicle parts or mats. This process can only be used on thermoplastics that soften when heated and harden again when cooled. More than 80% of plastics are of this type. The main drawback is that the melting temperature must be kept below 200°C, as polyvinyl chloride (PVC) decomposes at higher temperatures.

- *grinding down and mixing with fillers and adhesives*, followed by high pressure extrusion to produce 'low quality' plastics where finish and appearance are relatively unimportant. This process is used on thermoset plastics that are hardened by curing and cannot be re-melted or re-moulded

8.136   The calorific value of most plastics is high and this energy can be recovered. The options include direct incineration with energy recovery; the production of refuse derived fuel and packaging derived fuel for incineration; or use directly as a fuel. The production of refuse derived fuel or packaging derived fuel allows the plastic to be transported in a more economical manner, and leads to a more consistent feedstock for incineration. Significant opportunities exist to safely recover plastics in the manufacture of cement. Here, the energy content of the plastic is recovered at temperatures above 1,450°C and any resulting ash is incorporated in the product, thus avoiding landfill.

8.137   The packaging Regulations require a considerable increase in the amount of plastic waste recycled by 2001. Plastic must meet the 15% material-specific recycling target within the overall 50% recovery target for packaging waste (see Chapter 7 of this part of the strategy).

Recycling plastic bottles

Rigid plastic bottles and containers offer the greatest potential for increasing the level of recycling of household plastics packaging. RECOUP, an advisory body that promotes and facilitates plastic bottle recycling, operated a scheme in 1998, in conjunction with Rushmoor Borough Council. The aim of the scheme was to increase the number of plastic bottles collected in Rushmoor, to encourage best practice and to demonstrate to other local authorities that increases in plastic recycling could be achieved with little investment.

The *Message On A Bottle* competition was targeted at children and was well publicised in the local press, radio and television. A poster campaign was also run in schools, libraries, council offices and supermarkets. Householders were invited to write their name and telephone number on the plastic bottles they sent for recycling. A fortnightly draw then took place at the materials recycling facility and the winner was presented with prizes which included clothing such as fleece jackets, hats and sweaters made from recycled plastic bottles.

Reducing the volume of the bottles increases the economics of bottle recycling. Best practice was encouraged by making only bottles with their tops taken off and squashed eligible for a prize. The result was a 30% increase in the number of bottles recovered and an improvement in best practice.

8.138   Plastic wastes have generally been regarded as non-hazardous, although there has been lively public debate about PVC wastes, as there also has about PVC in general. Government policy on PVC wastes will be kept under review in the light of relevant studies undertaken. These include continuing UK life cycle assessments of PVC and a range of alternative materials, as well as EC Commission studies of the environmental impact of various waste management operations for PVC, which are expected to inform the development of a Commission proposal on PVC in 2000.

Power station ash

8.139   Pulverised fuel ash (PFA) and furnace bottom ash (FBA) are inevitable by-products of coal-fired electricity generation due to the incombustible material present in coal. The relevant totals for the UK during 1997 were 5.1 million tonnes of pulverised fuel ash and 1.4 million tonnes of furnace bottom ash.

8.140   The total amount of ash produced in the period 1995/96 has been reduced by almost 2 million tonnes since the previous period as a result of the changes in the coal mix. Over 46% of total ash was sold for re-use, with the remainder being sent to landfill. Almost all of the furnace bottom ash and over a third of the pulverised fuel ash was sold for re-use.

8.141   Pulverised fuel ash is used in a number of applications within the construction industry, including aerated concrete block manufacture, concrete products, structural fill, grouting of underground voids and as a partial replacement for cement in concrete. Furnace bottom ash is used primarily in the manufacture of lightweight blocks. Such use reduces the need for quarrying of aggregates with its associated environmental impacts. The amount of ash that can be sold is governed by the demand from the construction industry. Ash that remains unsold is disposed of in landfill sites and the total declined by over 1 million tonnes in 1996/97 compared with the previous year, due to the reduction in total ash production.

8.142   One problem facing the market for ash is the fact that the volume of work undertaken by the construction industry (which is one of the major customers) tends to vary in line with the wider economic performance of UK commerce and industry, which has a knock-on effect on the amount of ash which is re-used. Transport costs may also have a significant impact on the economic viability of re-use of ash, as power stations tend to be located in isolated areas away from potential end users.

8.143   Most of the power generating companies have in-house organisations that are responsible for developing markets for re-use of the ash. Examples include National Ash, a subsidiary of National Power, and Eastern Ash, a subsidiary of Eastern Group.

Blast furnace and steelmaking slags

8.144   Blast furnace slag is a by-product of the refining of iron ore. The refining process involves heating the ore in a furnace, which produces the metal in the form of pig iron and slag. The slag consists mainly of silicon, calcium and magnesium oxides.

8.145   The specific quantity of blast furnace slag produced depends on the raw materials used, but lies in the range 210 – 310 kilograms per tonne of pig iron produced. An estimated 4 million tonnes of blastfurnace slag was produced in the UK in 1996.

8.146   Several types of blast furnace slag can be produced depending on the manner and rate of cooling. If allowed to solidify naturally, a dense crystalline material, air-cooled slag, is produced. Slag can also be produced as granules and pellets by shock cooling the molten slag with water. Rapid cooling with air produces slag wool.

8.147   Air-cooled blast furnace slag is suitable for construction or road building. The actual demand for this material depends upon the construction activity local to the production sites. Granulated blast furnace slag is a higher value product that is used in the manufacturing of cement. To reduce the dependency on local markets for air-cooled slag, all four integrated steel works in the UK now operate granulation plants.

Slag produced and recycled in 1996		
	Produced (tonnes)	Recycled (tonnes)
Blast furnace slag	4,000,000	3,000,000
Basic oxygen steelmaking slag	1,500,000	200,000
Electric arc furnace slag	200,000	200,000

8.148   Steel slags result from the conversion of pig iron to steel. There are two types of steel slag; basic oxygen slag (BOS) and electric arc furnace (EAF) slag. Approximately 1.5 million tonnes of basic oxygen slag and 0.2 million tonnes of electric arc furnace slag were produced in 1996.

8.149   Steel slags can replace natural stone in a variety of applications. However, they contain free lime which hydrates causing a volume increase and therefore require weathering to allow expansion before use as a natural stone substitute. This represents a barrier to recycling steel slags.

British Steel plc

In order to increase the recycling rate of basic oxygen slag, British Steel are working with companies that market slag products to test the suitability of using basic oxygen slag over a range of applications. Possible uses being developed are as rail ballast and as a base material for road construction.

An environmental benefit of using slag as rail ballast is that it helps to avoid the degradation of the landscape caused by the quarrying of natural resources. Basic oxygen slag as ballast is hard, strong, resistant to weathering and stable.

Sewage sludge

8.150   Sewage sludge is an inevitable by-product of sewage treatment. It is produced at sewage works as a thick odorous liquid containing around 4% solid matter. About 35 million tonnes of raw sewage sludge is produced each year. Although it can be disposed in its raw form, sludge must be treated before it is used in agriculture. It is usually treated in a variety of ways to reduce odour, volume and, increasingly, pathogen levels prior to re-use or disposal. This treatment is undertaken at the sewage works and usually consists of one of the following methods:

- *dewatering* – the water content of the wet sludge is reduced by a mechanical process, such as by centrifuge, belt or filter pressing

- *thickening* – the concentration of the solids in the sludge is increased by using sun drying or gravity

- *digestion* – bacteria and micro-organisms are used to break down the organic matter in the sludge and produce methane gas for energy production

- *thermal drying* – high temperature drying techniques reduce the sludge to a granular form which is approximately 80% solids

- ***lime treatment*** – the addition of lime to lower the acidity for a specified length of time, which produces a material which is suitable for use as an agricultural liming agent

- ***composting*** – the biological treatment of sludge using composting techniques to produce a stabilised material with a high dry solids content

8.151   Treatment of the waste by the above methods reduces sludge production to approximately 25 million tonnes per year. Because the range of water contents is so wide, quantities are usually expressed in tonnes of dry solids. On this basis the quantity to be dealt with each year is around 1.1 million tonnes.

8.152   Sewage sludge production is expected to rise to 1.5 million tonnes per year by 2005 as a greater proportion of sewage is treated and higher treatment standards are applied under the phased implementation of the Urban Waste Water Treatment Directive. This Directive also prohibited the dumping of sewage sludge at sea by the end of 1998, which had previously been the means of disposal of around 30% of the total waste generation. The Directive also requires that sludge should be re-used whenever appropriate and that disposal routes should minimise the adverse effects on the environment.

8.153   For several decades the main option for re-use, accounting for around 50% of the total, has been the controlled application to agricultural land so that useful plant nutrients and organic matter are recycled to the soil. Sludge is applied to less than 1% of agricultural land. Other options for dealing with sludge include incineration and landfilling (8% each), land reclamation (6%) and use on dedicated sites (3%). Sludge previously dumped at sea now has to be disposed of by other routes such as spreading on agricultural land and energy recovery.

Examples of new technologies for the disposal or recovery of sewage sludges

Drying sludge by thermal treatment leads to a pasteurised, odour free product that can be easily stored for use when conditions are favourable. This allows water companies to have a greater flexibility in their sludge-to-land operations. Wessex water pioneered sludge drying technology in the UK by building a drier plant to take all of the sludge from the Avonmouth sewage works. The company promotes the dried granulated product (Biogran) as a soil conditioner and fertiliser. The current plant has a capacity of 13,000 tonnes of sludge per year.

Novel solutions to the challenge of sludge disposal include Northumbrian Water's sludge gasification scheme and Anglian Water's plans to divert sludge from sea disposal into the manufacture of building materials.

Northumbrian Water is building a 33,000 tonnes per year sludge drying project at its Bran Sands site at Teesside. The facility will handle sludges previously dumped at sea from Teesside, Tyneside and Wearside. The dried sludge will be gasified in an £11 million plant currently being built at Bran Sands. The plant is the first major gasification scheme for energy generation in the UK and will produce 5MW of electricity. Drying and gasification capacity will be expanded to deal with the company's expected 50% increase in sludge production by 2005.

Before the end of 1998, Anglian Water dumped some 11,000 tonnes of sludge each year at sea from Tilbury, but this is now diverted to a plant making building materials. The project is a joint venture with Tilbury Aggregates. The sludge is mixed with clay and pulverised fuel ash and "cooked". The sludge expands to make a lightweight material suitable for blocks and bricks.

8.154 The Government and the National Assembly consider that recovering value from sludge through spreading on agricultural land is the Best Practicable Environmental Option for sludge in most circumstances. Agricultural use of sludge brings savings of several million pounds in fertiliser costs and the organic matter improves soil structure, its workability and its water holding capacity. The Sludge (Use in Agriculture) Regulations 1989 controls the spreading of sludge on agricultural land. The main aim of the controls is to prevent the accumulation of hazardous concentrations of heavy metals in soil and to prevent bacteriological contamination of crops. The regulations are augmented by a Code of Practice that specifies the way sludge should be treated, utilisation practices on farms and monitoring requirements.

8.155 Significant investments are currently being made by water companies to deal with the end of sea dumping and the increased quantities of sewage sludge. Water company strategies are based on finding secure outlets, in particular by greater use of agricultural land and in some cases partly on increased incineration. Recent predictions suggest that by 2005 the proportions used on land, incinerated or gasified, or landfilled will be 60%, 36% and 4% respectively, but this depends on continued confidence in the agricultural route and other developments such as new constraints on landfilling sludge.

INITIATIVES ON SEWAGE SLUDGE

8.156 To maintain and improve confidence in the agricultural use of sewage sludge, the Government and the National Assembly will ensure that appropriate and precautionary controls are in place which have the support of the public, the farming and food industries, and other experts.

8.157 The Government commissioned a comprehensive review of the scientific evidence underpinning the current controls on the agricultural use of sewage sludge, following the recommendations of the Royal Commission of Environmental Pollution in its report *Sustainable Use of Soil*. The review, undertaken by the Water Research Centre and completed in July 1998, takes into account new evidence and, in particular, concerns about new pathogens. It makes recommendations for changes in practice and for further research.

8.158 The Government announced its acceptance of the main recommendations in its response to the report of the Select Committee on the Environment, Transport and Regional Affairs on *Sewage Treatment and Disposal*. In parallel with the review there have been extensive discussions involving the water industry, major food retailers, the Environment Agency and Government Departments to agree the necessary measures to ensure the long-term confidence of all parties in this valuable method of waste recycling.

8.159 The Government and the National Assembly support the outcome of the scientific review and the discussions with stakeholders. The resulting *Safe Sludge Matrix* (an informal agreement between the water industry and the major food retailers) has resulted in a range of precautionary changes in the requirements for the use of sewage sludge on agricultural land. The significant changes are:

- phasing out all use of untreated sewage sludge on agricultural land by the end of 2001. Earlier phase-out dates apply to particular uses of untreated sewage sludge, including the end of 1998 for use on grass for silage and grazing and maize for silage, and the end of 1999 for certain combinable crops and animal feed crops

- more stringent requirements for the performance of sludge treatment processes, with a distinction drawn between conventional treatment and enhanced treatment, and the introduction of performance monitoring and auditing provisions

- phasing out the surface application of conventionally treated sludge to grass for grazing by the end of 1998. Subsequent applications must be injected unless the sludge has been subject to advanced-level treatment

- stricter post-application controls when conventionally-treated sludge is used, including an interval of 12 months between application and harvest of field vegetables and 30 months where the vegetables are eaten raw

- reduction of the maximum concentration of lead in soil from 300 to 200mg/kg as a precautionary measure to limit metal accumulation in animal offal under exceptional circumstances

- The action programme under the Nitrate Vulnerable Zones (England and Wales) Regulations 1998 restricts the application of all fertilisers including sewage sludge in designated areas

8.160   The Government intends to make the necessary amendments to the statutory framework of controls and the associated Code of Practice. In Wales this will be a matter for the National Assembly. Water companies are to revise their sludge disposal strategies to take account of the new requirements and make them available for scrutiny by the Environment Agency. In scrutinising companies' sludge strategies, the Government expects the Agency to ensure that companies comply with the requirements of the new regulations within the terms of the proposals set out above; and that, within this constraint, maximum reasonable use should continue to be made of the agricultural recycling route.

Contaminated soil

8.161   The construction and demolition industries in England and Wales consigned over 1 million tonnes of special waste in 1997/8. The largest proportion of this waste consisted of around 500,000 tonnes of contaminated soils (20% of the total consigned special waste for that year). It is believed that this figure may actually overestimate the quantities of waste soils that are special waste. This is because it is often more practical to consign all wastes arising from a contaminated site as special rather than carry out the complex analysis required to separate special from non-special waste soils.

8.162   More than 90% of the contaminated soil consigned in England and Wales in 1997/8 was consigned to landfill. It is widely recognised that there will need to be a reduction in the landfilling of contaminated soils, and this may be further driven by the implementation of the Landfill Directive. Research and development of soil treatment and cleaning technologies will increasingly enable soils to be recycled for beneficial use, break down contaminants or extract them for further treatment and/or disposal. It is expected that there will be increasing use of such technologies to recover soils and remove contaminants both on and off site. These issues are dealt with in the following guidance on contaminated land redevelopment, treatment and remediation:

- CIRIA (1995 & 1996) *Remedial Treatment of Contaminated Land.* Reports SP101-111. Published by Construction Industry Research and Information Association, 6 Storey's Gate, London SW1P 3AU

- DETR (1998) *International Review of the State of the Art in Contaminated Land Treatment Technology Research and a Framework for Treatment Process Technology Research in the UK.* Published by the Department of the Environment, Transport, and the Regions, London. ISBN 1-85112-117-X

- Martin and Bardos (1996) *A Review of Full Scale Treatment Technologies for the Remediation of Contaminated Soil.* Report for the Royal Commission on Environmental Pollution. Published by EPP Publications, Richmond, Surrey. ISBN 1-900995-00-X

8.163   The increasing use of the treatment and recovery of contaminated soils may lead to a reduction in the quantities of waste soils being consigned (in accordance with the requirements of the Special Waste Regulations), although some of these processes may result in residues that constitute smaller quantities of more hazardous wastes.

8.164   On the other hand, a number of factors, such as increased redevelopment of brownfield sites and the implementation of the new contaminated land regulations, may result in the redevelopment and remediation of more contaminated sites. This could lead to an increase in the quantity of waste contaminated soils being generated.

Textiles and clothing

8.165   Estimates for the generation of textile waste vary between 550,000 tonnes and 900,000 tonnes each year. Post industrial waste textiles arise during yarn and fabric manufacture, garment making processes and from the retail industry.

8.166   The Textile Recycling Association estimates that 25% of discarded post-consumer textiles are currently recovered. Of these recovered textiles, 43% becomes second-hand clothing, 12% wiping cloths, 22% filling materials, 7% goes for fibre reclamation, 9% are second-hand shoes which are reused and 7% is rejected. Textiles made from both natural and man-made fibres can be recycled.

8.167   A significant proportion of post-consumer textiles is collected by charities for re-sale as re-wearable quality clothing. If the quality of the recovered items is too poor, recycling can be the next best option. In terms of environmental benefits, textile recovery:

- reduces demand for virgin resources
- results in less pollution and energy savings as fibres do not have to be transported from abroad

8.168   Recycling fibre avoids many of the polluting and energy intensive processes needed to make textiles from virgin materials, including:

- savings on energy consumption when processing, as items do not need to be re-dyed or scoured

- less effluent, as raw wool has to be thoroughly washed using large volumes of water

- reduction of demand for dyes and fixing agents and the problems caused by their use and manufacture

8.169   Textiles that are not re-used or recycled are either incinerated with household waste or sent to landfills for final disposal. It is estimated that in the UK 400,000 – 700,000 tonnes of textiles are landfilled each year. Prices for both waste textiles from clothing banks and sorted textiles increased significantly in mid 1996. Since 1997, however, prices have dropped, and are now similar or below their 1995 values. This may have a detrimental effect on textile recovery.

MANAGEMENT OF TEXTILE AND CLOTHING WASTES

8.170   At present the public has the option of putting textiles in clothes banks, taking them to charity shops or having them picked up from their homes. Recyclatex, a scheme run by the Textile Recycling Association in conjunction with local authorities, provides textile banks for public use. The Salvation Army, Scope, and Oxfam also use a bank scheme in conjunction with other methods. The Salvation Army, for example, combines the use of over 1,500 banks with door-to-door collections and these account for the processing of in excess of 200 tonnes of clothing each week.

8.171   All collected textiles are sorted and graded by experienced workers who are able to recognise the large variety of fibre types resulting from the introduction of synthetics and blended fibre fabrics. Once sorted, the items are sent to various destinations according to their quality. For instance, re-wearable textiles (including shoes and clothes) can often be resold in the UK and abroad. Unwearable textiles can be re-used and recycled as follows:

- woollen garments can be sold to specialist firms for fibre reclamation to make yarn or fabric

- trousers, skirts and other garments can be sold to the flocking industry – items are shredded for fillers in car insulation, roofing felts, loudspeaker cones, panel linings, furniture padding

- cotton and silk can be sorted into grades to make wiping clothes for a range of industries from automotive to mining, and for use in paper manufacture

8.172   Post industrial waste is often reprocessed in-house. Clippings from garment manufacture are also used by fibre reclaimers to make into garments, felt and blankets. Mills grade incoming material into type and colour. The colour sorting means no re-dying has to take place, saving energy and pollutants. Initially the material is shredded into 'shoddy' (fibres). Depending on the end uses of the yarn other fibres are chosen to be blended with the shoddy, and the blended mixture is spun ready for weaving or knitting.

> ### Evergreen Recycled Fashions
>
> Evergreen Recycled Fashions is one of a few remaining European textile regenerators, turning worn sweaters and rags into new yarns and fabrics by a series of processes which begin with pulling the garments apart in a machine with counter-rotating spiked rollers. This restores the garments to their constituent fibre. Buttons are mechanically removed at this stage, while zips must be removed by hand in advance.
>
> To avoid the need for re-dyeing – a process that has high water and chemical demand, and generates large quantities of effluents needing treatment – material is hand sorted into 40 standard colours and shades. A series of mechanical processes then turns the fibres into a spun yarn for sale either as hand-knitting yarn or for the manufacture of cloth for making up into new garments. Evergreen uses both high wool content fibre mixtures and wool/man-made mixtures.
>
> Evergreen claims that if every Briton purchased one reclaimed woollen garment each year – which would far exceed the company's present small output – the resulting savings would be:
>
> - 371 million gallons of water
> - 480 tonnes of chemical dyes
> - 4517 million days of an average family's electricity needs
>
> Evergreen's re-use of fibres avoids the processes involved in scouring raw new wool to remove dirt and animal secretions which are energy and water intensive, as well as producing polluting effluents. The above calculations do not include savings in imports, nor environmental costs of landfilling discarded textiles, so (Evergreen argues) the total environmental benefits of textile regeneration are even greater.

8.173   Changes in the management of waste textiles currently consists of process changes or the introduction of new operational or management methods and not necessarily new technology.

8.174   An example of a process change is to switch from a manual handling system to an automated feeding system, which can quadruple a reclamation company's throughput without any expansion of premises or an increase in staffing levels.

8.175   An example of an operational change is the storage of *out-of-season* clothes by Oxfam. One of the problems faced by charities such as Oxfam is the fact that people tend to donate winter clothing in the spring and summer clothing in the autumn and winter. Following sorting at clothing depots, Oxfam will send clothes suitable for the season to its shops, whereas clothes which are out-of-season will be stored and only sent to stores at the appropriate time of year. This ensures that a greater percentage of donated clothes are sold for re-use, reducing the quantities that are sent for processing and recycling.

Tyres

8.176   In January 1998, the Used (formerly Scrap) Tyre Working Group (UTWG) was asked by the Government to recommend its preferred means of ensuring that the UK would be able to meet the virtual 100% recovery rate for scrap tyres, implied by the Landfill Directive, on time and at least cost. This Directive will ban the landfilling of both whole tyres and shredded tyres.

8.177   The Used Tyre Working Group report covered used tyre disposal statistics for 1998. This estimated that some 40 million tyres (465,000 tonnes) were scrapped in 1998, of which value was recovered from approximately 70%. Retreading accounted for 18%, energy recovery 18%, re-use 18%, recycling 10% and landfill engineering 5%. The report showed that a number of mechanisms for ensuring compliance with the landfill ban had been examined including statutory producer responsibility and levy arrangements. However, the Group favoured a market-based approach, although it recognised that it was not yet certain that this approach would guarantee the 100% requirement was met. A forecast of developments to 2003 showed an upward trend in the recovery rate over the next four years, with the growth in the use of tyre derived fuel in cement kilns being a particularly notable feature. The forecast anticipated a 90% scrap tyre recovery rate in 2003. There would then be a further 3 years, leading up to 2006, to bridge the remaining 10% gap.

8.178   The Government response, through DTI Ministers, agreed with the industry that it was not yet necessary to take a decision on whether the UK would be in a position to meet the 100% recovery requirement (in 2003 for whole tyres and 2006 for shredded tyres) to landfill sites, through market measures alone. However, due to the lead time needed for statutory measures to be brought into force and have effect, a decision on whether to press forward with statutory measures would be required significantly before the landfill ban came into force. In the meantime, work will continue on developing statutory measures to the point where they could be rapidly implemented. This time will also be used to further develop understanding of likely developments in the scrap tyre market over the years leading up to the landfill ban, which would then help guide decisions.

8.179   The Used Tyre Working Group is formed of the four main tyre trade associations, the British Rubber Manufacturers' Association, the Imported Tyre Manufacturers' Association, the National Tyre Distributors Association and the Retread Manufacturers Association, together with officials from the Department of Trade and Industry.

Waste wood

8.180   Wood is an ubiquitous material which appears in a number of waste streams, including construction and demolition wastes, agricultural wastes, municipal wastes, bulky household wastes (including fittings and furniture), packaging waste, and other commercial and industrial wastes.

8.181   However, in general terms wood wastes can be thought of as arising in the following forms:

- green wastes, as a by-product from the management of trees in urban, suburban and rural areas; and the maintenance of both natural and commercially planted forests

- untreated wood and timber – packaging waste made from untreated wood, for example pallets, can also be included in this category

- structural wood waste, which will probably have been treated with preservative and other chemicals – this includes old telephone poles and railway sleepers, as well as wood used in commercial, municipal and domestic buildings, and in the construction and demolition of those buildings and other infrastructure

- waste from wood processing, sawdust, sanding dust, shavings, bark and off-cuts

- waste manufactured products, some of which will have been made entirely from wood, and others which will include wooden components – products range from car fittings, to furniture, toys, commercial and domestic utensils, detachable commercial and domestic fixtures and fittings

8.182   Waste paper and cardboard, although derived in part from wood pulp, is considered to be a separate waste stream. Reconstituted wooden panels, such as MDF or chipboard, are derived from the recycling of waste wood.

EXEMPTIONS FOR RECOVERY OF WASTE WOOD

8.183   A number of exemptions from waste management licensing are in place to encourage the recovery of waste wood, providing that certain requirements are fulfilled. General rules must be adopted for each type of activity laying down the types and quantities of waste and the conditions under which the activity in question is exempted. In addition, the activities must ensure that the waste is recovered without harming human health or the environment. Exempted activities include the manufacture of products and finished goods from waste wood as well as chipping and shredding for the purposes of recovery and re-use.

8.184   There is an exemption from the requirement to have a waste management licence for the burning of waste wood on land in the open in specific places, and provided it is only burned on the land where it is produced. Garden bonfires are not subject to the provisions of the waste management licensing system that applies to establishments and undertakings, and does not include private individuals.

SUSTAINABLE WASTE MANAGEMENT OF WASTE WOOD

8.185   Managing wood wastes will be determined largely by which group the waste wood arises from. Untreated wood – either from tree management or from structural wood waste – will be easier to manage in a more sustainable manner than wood contaminated with preservatives or paints, which may in some circumstances be classified as special wastes. However, the use of preservatives to treat timber may reduce the quantity of wood wastes produced, as the lifetime of the treated products may be 10 or 20 times those of untreated products.

8.186   For manufactured products, the percentage of wood components in a product will affect its treatment, and consideration will need to be given to whether the product as a whole can be refurbished, or if the wood component can be salvaged from the rest of the product. Often, refurbishment of a product will involve replacing the wooden portion of the product.

8.187   The whole range of waste management options can be applied to wood waste, although for treated timber some forms of management will be inappropriate depending upon the type of treatment. For example, wood treated with copper chrome arsenic (CCA) should not be burned or incinerated, and wood treated with tributyl tin oxide (TBTO) or tributyl naphthenate (TBTN) should not be used (or re-used) in situations where it may come into contact with surface waters.

8.188   Virgin wood from sustainably managed forests can often be a more sustainable option than other virgin materials such as metals, or oils and chemicals for plastics. Even so, there will often be scope for using recycled or refurbished wood in the structure or the manufacturing process, in place of virgin wood. For example, wooden telegraph and electricity poles are routinely given away or sold for re-use in non-structural applications such as fencing.

8.189   Wooden products are frequently *re-used* or *refurbished*. Indeed a significant proportion of today's antique and second hand market is based on the resale of refurbished wooden furniture and household fittings. A classic example of re-use is the use of wooden pallets and tea chests for packaging and transporting goods.

8.190   Untreated wood is also routinely *recycled*: after chipping and pulping, waste wood is used in the manufacture of paper, cardboard, chipboard and MDF. Using recycled wood in these processes is preferable to using virgin wood, and there will be no difference in the quality of the recycled and virgin material. Treated wood, or wood contaminated with hazardous chemicals, should not be used in feedstock.

8.191   *Composting* of waste wood (once it has been chipped) is a viable option for households with gardens, and also for local authorities that have a significant number of trees to manage. *Landspreading* of chipped wood waste is a viable recovery option if it results in benefit to agriculture and ecological improvement. The addition of wood waste can act to improve poor soil quality. Treated wood is not suitable for composting as it will be slow to break down and may contain hazardous residues.

8.192   *Incineration with energy recovery* will often be the best practicable environmental option for wood in some circumstances, including particularly waste that has been either treated with preservatives or has been contaminated with other potentially hazardous chemicals (though waste wood treated with copper chrome arsenic (CCA) should not be burned or incinerated). *Bonfires* of wood and other wastes are a common occurrence during summer and autumn (as are municipally arranged bonfires in November and at other times of communal celebrations). Unnecessary bonfires should be discouraged as they are a wasteful use of wood, and a significant source of dioxin emissions to the environment. Burning treated wood, especially CCA treated wood, can result in significant concentrations of pollutants in smoke and may result in ashes containing significant concentrations of hazardous substances. Burning wood in tightly controlled energy from waste plants has the advantage over bonfires in that by-products of the process such as dioxins can be scrubbed from the smoke, and the resulting emission to the atmosphere can be closely monitored.

8.193   Significant quantities of waste wood are still disposed of to landfill. Landfill rarely represents the best practicable environmental option for waste wood, and should be discouraged where possible. Waste wood is biodegradable, thus wood arising in municipal waste streams will be affected by the landfill diversion targets agreed for biodegradable wastes in Article 5 of the Landfill Directive. For waste streams not affected by Article 5 other measures – including education and information campaigns for businesses – will need to be considered if waste wood is to be moved away from landfill to more sustainable waste management options.

ANNEX A

Major waste facilities in England and Wales

A.1   Decisions on the planning for, and location of, waste management facilities should (for the most part) be taken at the local level. Chapter 3 of this part of the strategy gives details of the operation of the waste planning system, and the role of waste development plans within that system.

A.2   Local planning authorities are required to draw up development plans under Town and Country Planning legislation. Waste local plans, or, in metropolitan areas and Wales, the waste aspects of unitary development plans, are required to consider where waste management facilities may be located in the context of overall development of the area; and also to set out the amenity and land use criteria that should be applied to proposals for the development of waste facilities.

A.3   Under the Town and Country Planning legislation, planning authorities must have regard to national and regional policies, including policies on waste management, in drawing up their waste development plans. **This waste strategy will be a material consideration for planning authorities in drawing up their development plans and for determinimg individual planning applications**.

A.4   The following maps indicate the locations of various types of waste management facilities across the UK.

Map 1
UK Glass Reprocessors
and Cullet Processors

★ Glass Reprocessors

◆ Cullet Processors

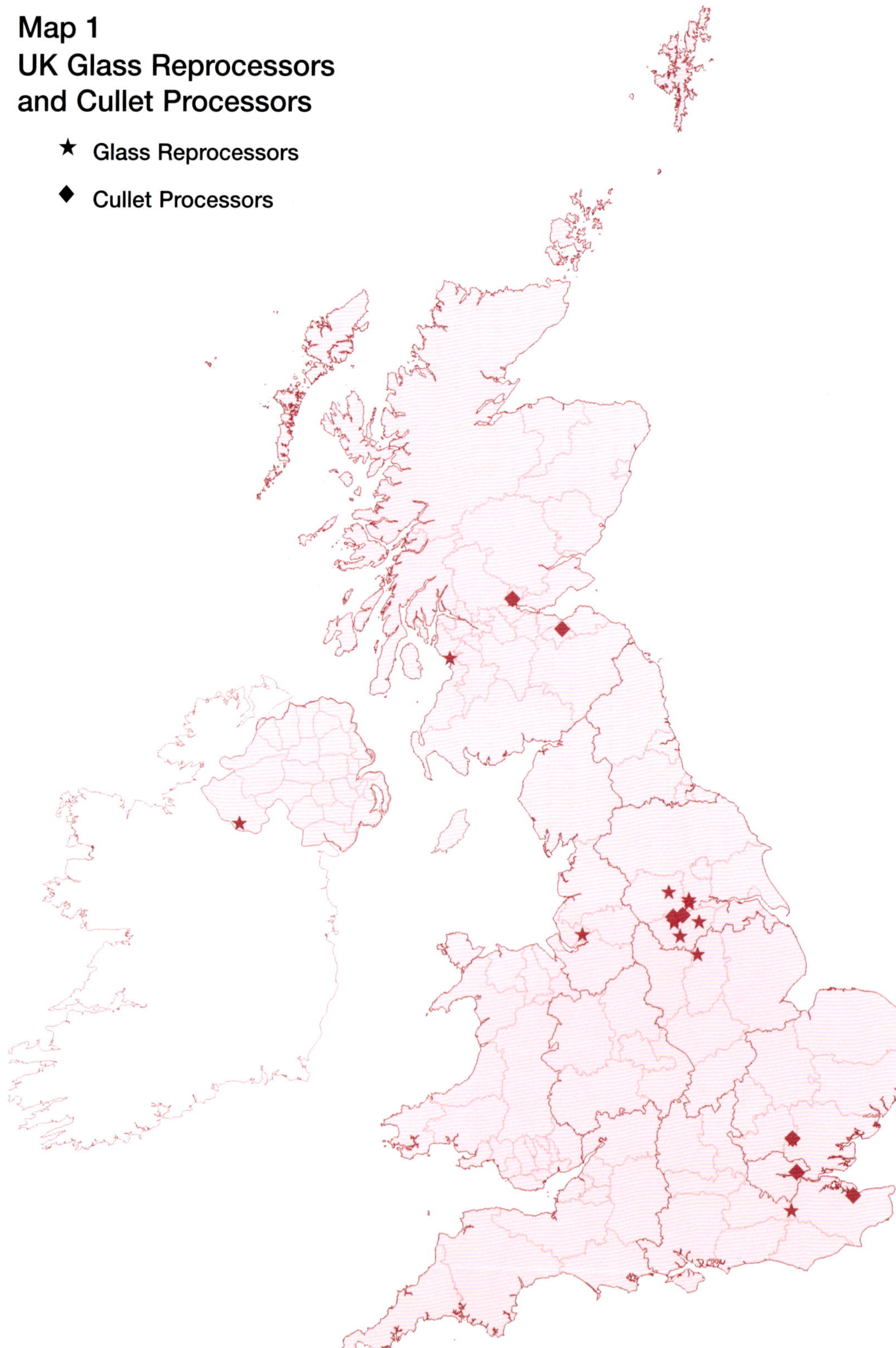

Map 2
UK Metals Processors (Fragmentisers)
– Heavy and Light Scrap

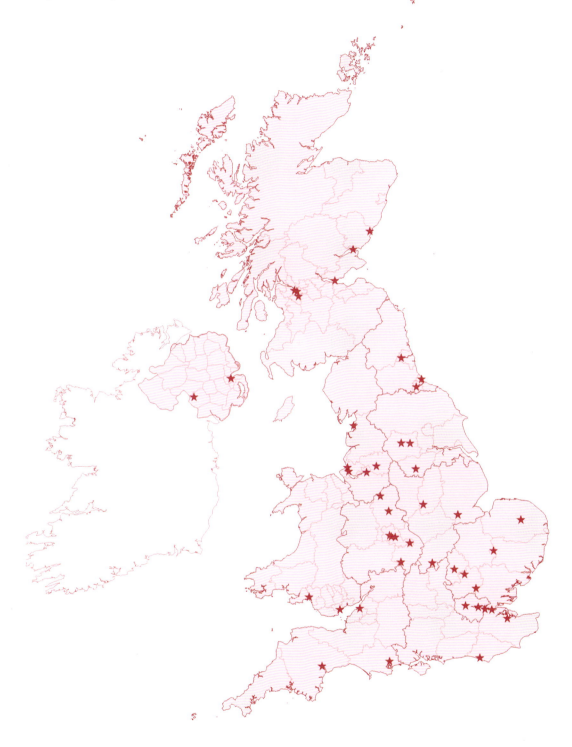

Map 3
UK Steel Reprocessors
(Steel Mills)

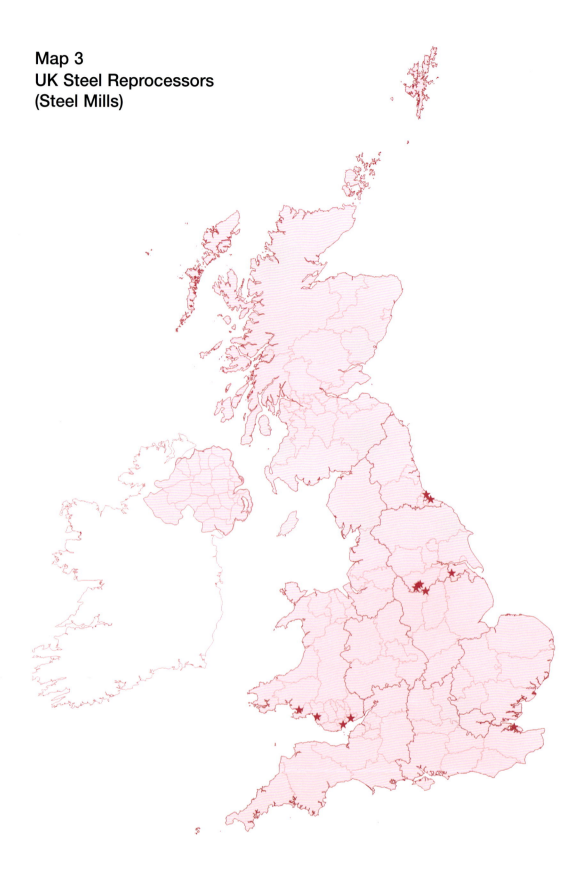

Map 4
Large UK Aluminium Reprocessors and Regional Aggregation Centres

★ Aluminium Reprocessors

◆ Regional Aggregation Centres

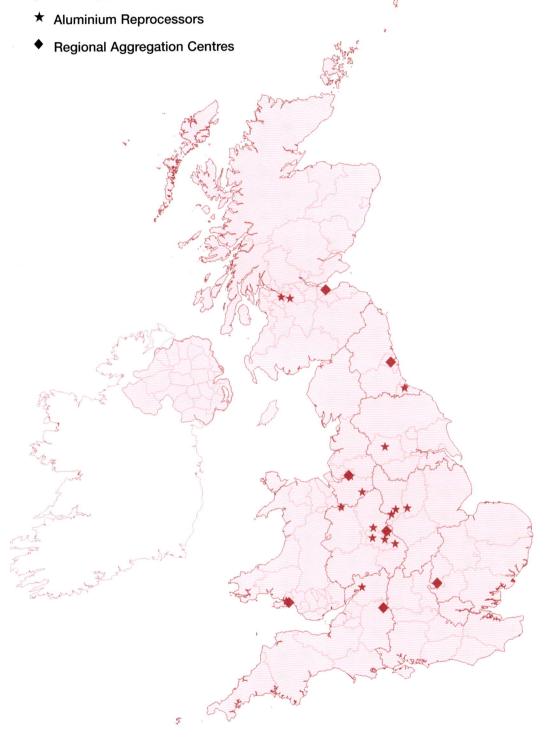

Map 5
UK Paper Processors
(Baling Operators)

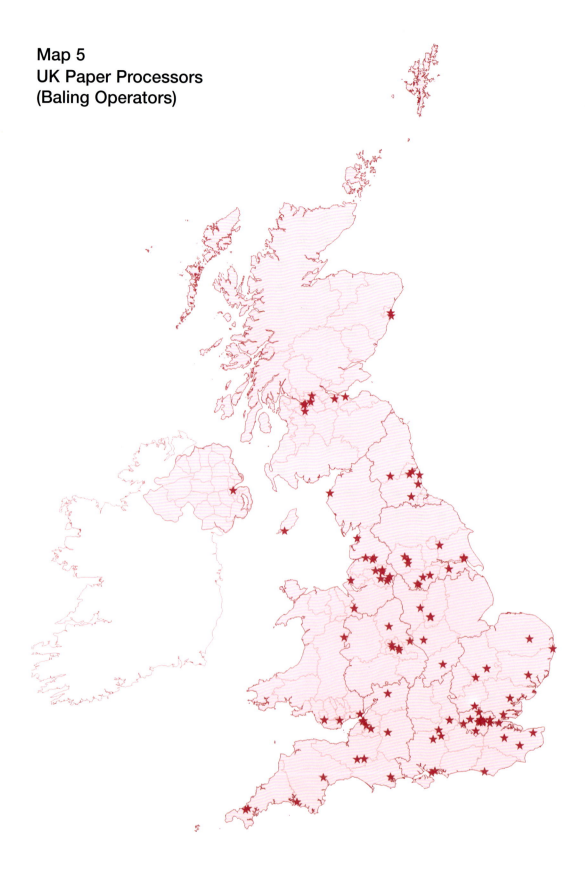

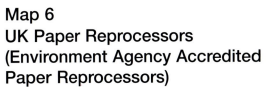

Map 6
UK Paper Reprocessors
(Environment Agency Accredited
Paper Reprocessors)

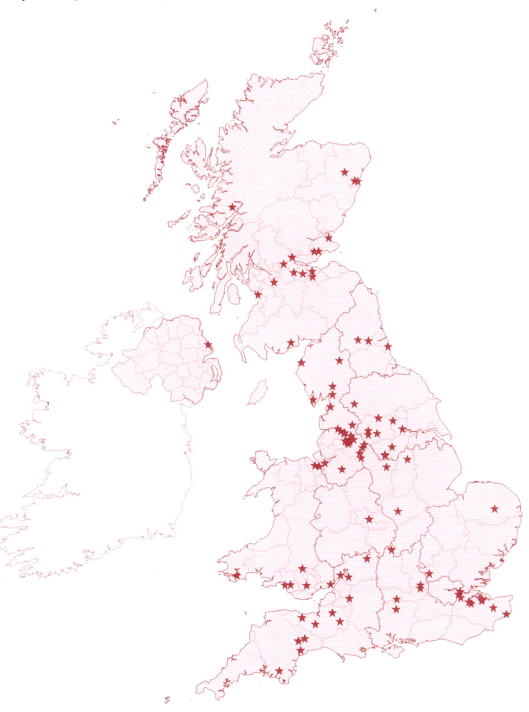

Map 7
UK Plastic Reprocessors
(Granulators and Melting Operators)

★ Melt Process Sites

■ Granulation Shredding Only Sites

▲ Unknown

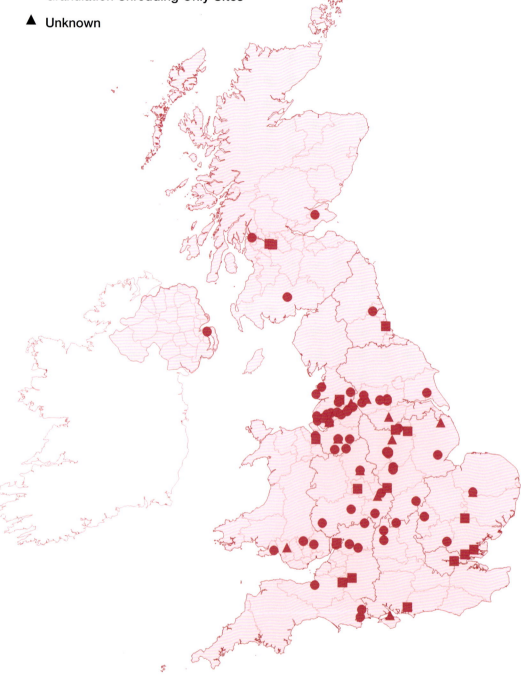

Map 8
Textile Reclamation Sites

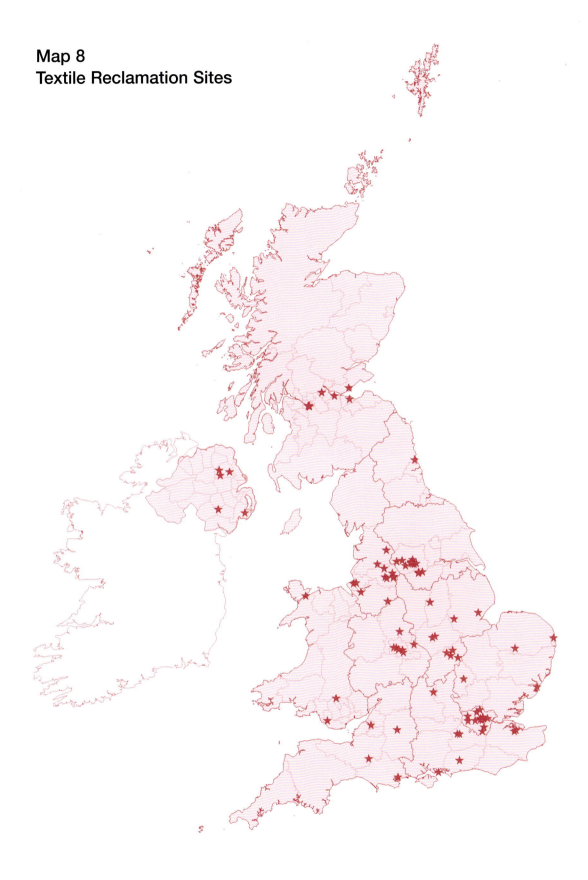

**Map 9
Major UK Wood Processors
(Wood Chippers)**

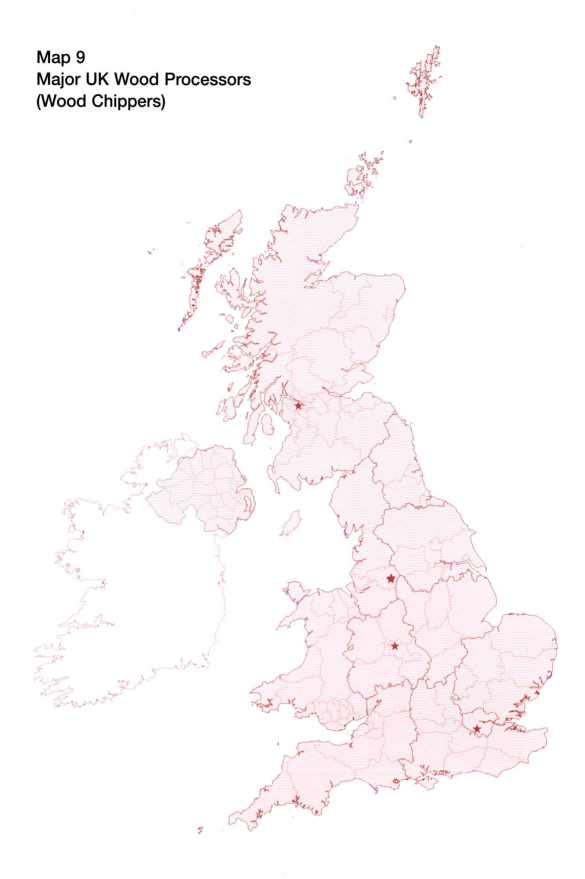

Map 10
Major UK Wood Reprocessors
(Panelboard manufacturers)

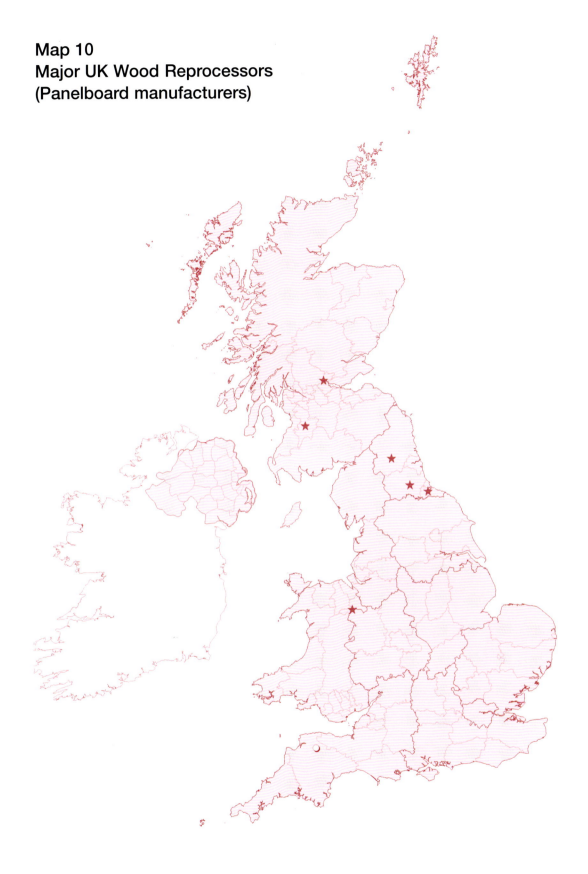

Map 11
UK Motor Oil Reclamation Sites

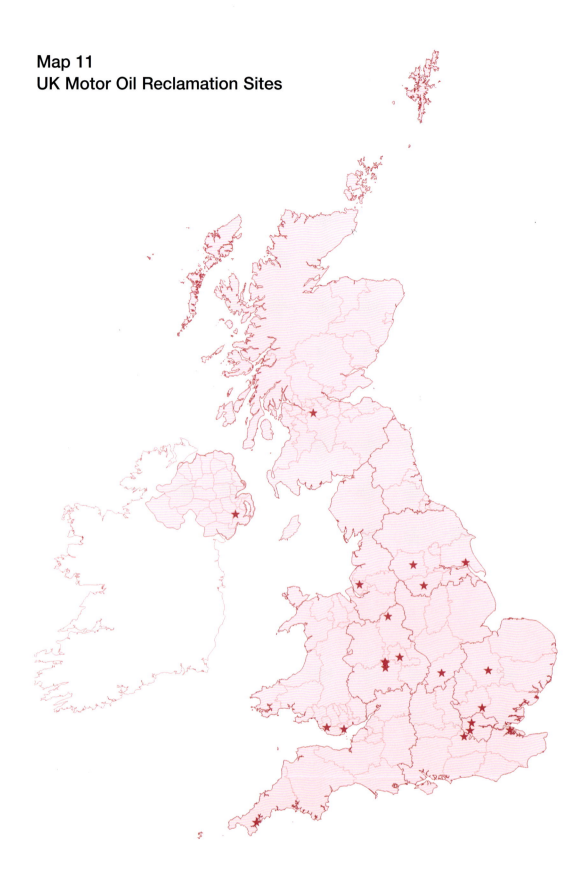

Map 12
UK Catering Oil Reclamation Sites

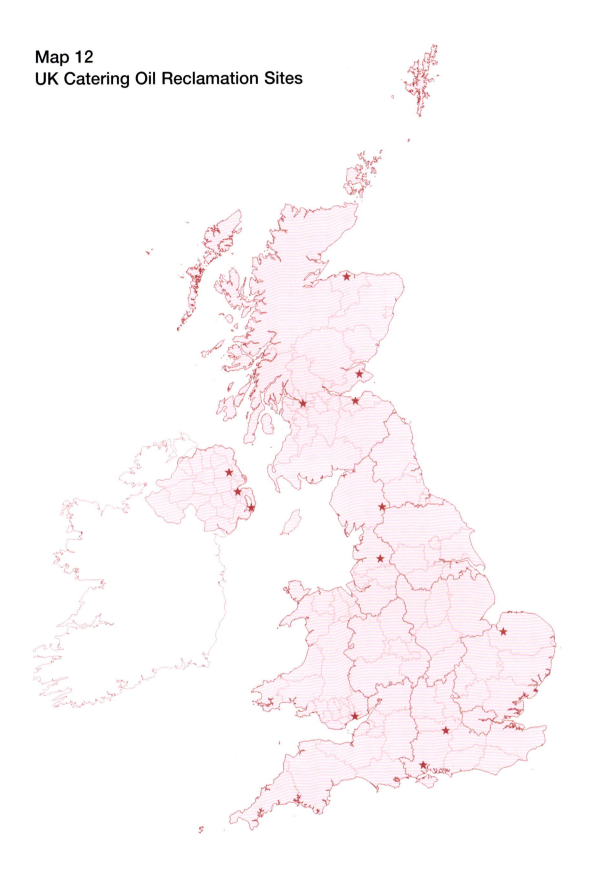

**Map 13
UK Energy from
Waste – Mass Burn
Combustion Sites
(1998)**

♦ Approved yet not constructed

★ Operational

● Proposed

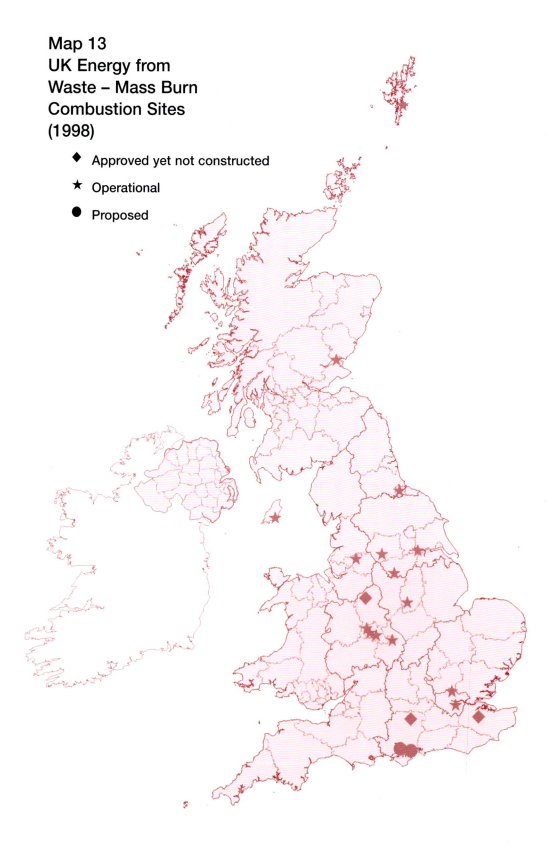

Map 14
UK Energy from
Waste – Tyres and
Secondary Liquid
Fuels (1998)

▼ Secondary Liquid Files

■ Tyres

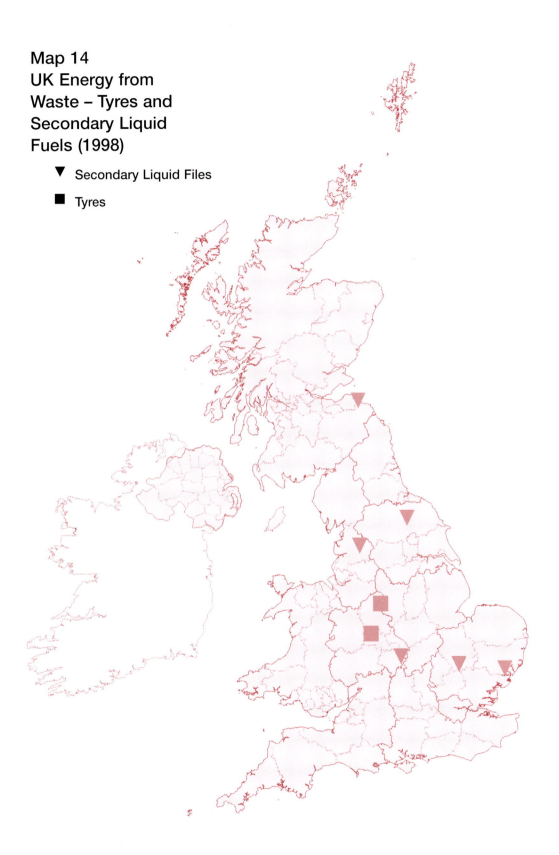

**Map 15
UK Energy from
Waste – Refuse
Derived Fuel
Combustion Sites
(1998)**

▲ RDF

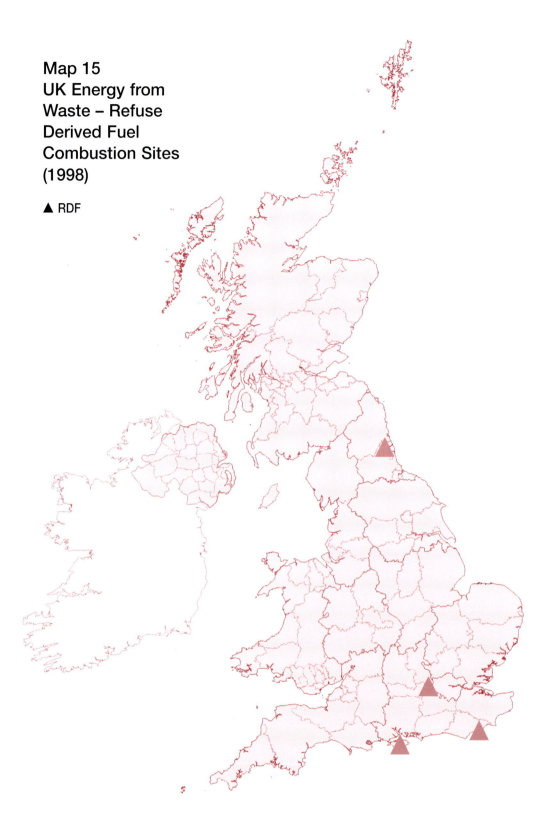

Map 16
UK Composting Sites

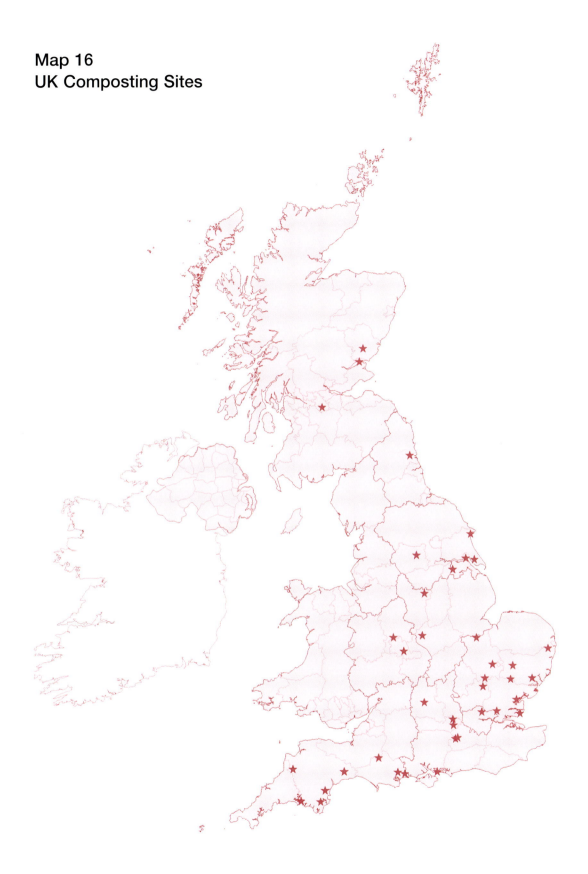

**Map 17
UK Ozone Depleting Substances
Recovery Sites**

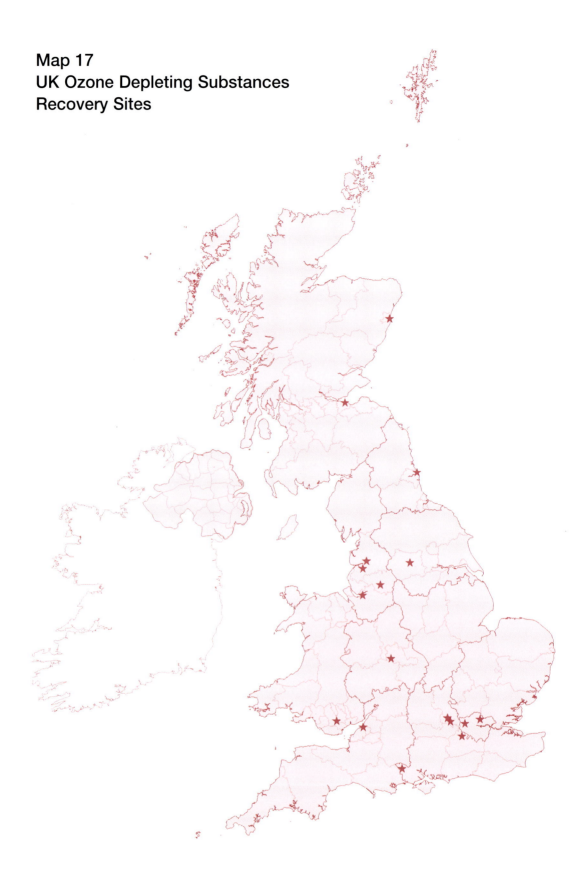

Map 18
UK Materials Recycling Facilities Sites (1998)

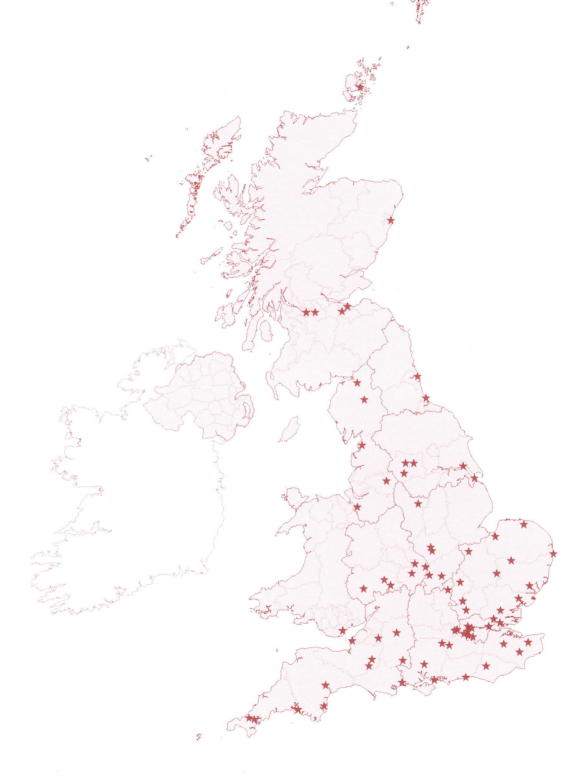

Map 19
UK Construction and Demolition Sites

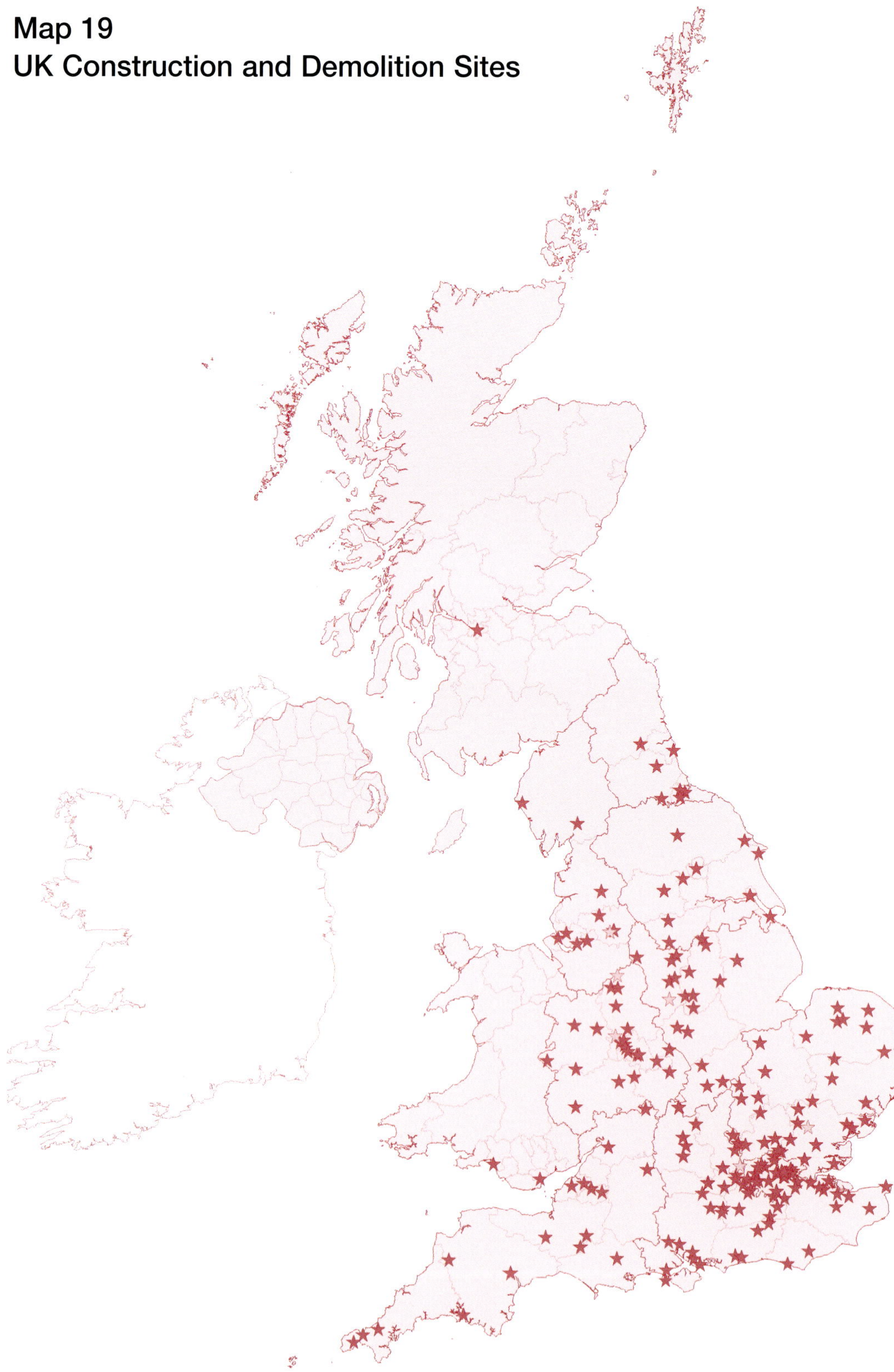

ANNEX B

Legal framework for waste management

B.1   This annex identifies the current (and developing) legal framework within which waste management operations take place. The annex is divided into three sections:

- international and European legislation
- legislation in England and Wales
- the legal definition of waste

International and European legislation

B.2   Waste legislation in England and Wales is driven by our need to manage our waste safely and effectively, and also by our international commitments and undertakings, particularly in the European Union.

B.3   The principal Directive controlling waste management throughout the European Union is the Framework Directive on Waste (Council Directive 75/442/EEC as amended by Council Directive 91/156/EEC and adapted by Council Directive 96/350/EC). The provisions contained in the Framework Directive are implemented into law by the Environmental Protection Act 1990, as amended by the Environment Act 1995, together with a number of Regulations on various aspects of waste management.

B.4   This strategy, together with guidance to planning authorities, implements for England and Wales the requirements within the Framework Directive on Waste, the Hazardous Waste Directive (Council Directive 91/689/EEC) and the Packaging and Packaging Waste Directive (European Parliament and Council Directive 94/62/EC) to produce waste management plans. It is also a strategy for dealing with waste diverted from landfill in England and Wales, as required by the Landfill Directive (Council Directive 1999/31/EC).

B.5   Playing a leading and influential role in the development and improvement of international and European law is a key aim for the Government and the National Assembly. European legislation provides a clear framework which applies across the Single Market; and the Basel Convention, which regulates and aims to reduce international trade in waste, helps to ensure that countries cannot simply export their waste for disposal (see Chapter 2 section 2.7 of this part of the strategy).

European Union legislation and actions

B.6 The legislative framework for sustainable waste management is an area that is continually evolving as we learn more about the processes and effects (both short-term and longer-term) of various waste management options. This section lists current (2000/01) European legislation related to waste, and then considers possible developments at the European level.

CURRENT EUROPEAN WASTE AND WASTE-RELATED LEGISLATION

B.7 These selected European Decisions and Directives either direct or influence waste policy in England and Wales:

- Directive 75/439/EEC – Disposal of waste oils
- Directive 75/442/EEC – Framework Directive or Waste (as amended by 91/156/EEC)
- Directive 76/403/EEC – Disposal of PCBs and PCTs
- Directive 78/176/EEC – Titanium dioxide industry waste
- Directive 78/319/EEC – Toxic and dangerous waste
- Directive 80/68/EEC – Groundwater
- Directive 82/883/EEC – Titanium dioxide pollution
- Directive 85/337/EEC – Environmental impact assessment
- Directive 86/278/EEC – Sewage sludge
- Directive 87/101/EEC – Disposal of waste oils
- Directive 89/369/EEC – Prevention of air pollution from waste incinerators
- Directive 89/429/EEC – Prevention of air pollution from waste incinerators
- Directive 90/425/EEC – Animal waste
- Directive 90/667/EEC – Animal waste
- Directive 91/157/EEC – Batteries and accumulators
- Directive 91/271/EEC – Urban waste water treatment
- Directive 91/689/EEC – Hazardous waste
- Directive 91/692/EEC – Standardising and rationalising reports on the implementation of certain environmental directives
- Directive 92/112/EEC – Titanium dioxide waste pollution
- Directive 93/86/EEC – Batteries and accumulators
- Directive 94/31/EC – Hazardous waste
- Directive 94/62/EC – Packaging and packaging waste
- Directive 94/67/EC – Hazardous waste incineration
- Directive 96/59/EC – Disposal of PCBs and PCTs
- Directive 96/61/EC – Integrated pollution prevention and control
- Directive 97/11/EC – Environmental impact assessment
- Directive 1999/31/EC – Landfill
- Decision 93/98/EEC – Control of transboundary movements of hazardous waste and their disposal (Basle Convention)
- Decision 94/3/EC – European waste catalogue
- Decision 94/575/EC – Supervision and control of shipments of waste
- Decision 94/774/EC – Supervision and control of shipments of waste
- Decision 94/904/EC – Hazardous waste list
- Decision 96/350/EC – Waste
- Decision 96/660/EC – Supervision and control of shipments of waste
- Decision 97/129/EC – Packaging and packaging waste

- Decision 97/138/EC – Packaging and packaging waste
- Recommendation 81/972/EEC – Re-use of waste paper and use of recycled paper
- Regulation (EEC) No 259/93 – Supervision and control of shipments of waste
- Regulation (EC) No 120/97 – Supervision and control of shipments of waste

B.8 Selected European Waste Policies include:

- European Waste Strategy (Commission communication to the Council and the European Parliament of 18 September 1989)

- Review of European Waste Strategy (Commission report to the European parliament and to the Council on waste management policy, 8 November 1995)

- Fifth Environmental Action Programme

DEVELOPING WASTE POLICY AT THE EUROPEAN LEVEL

B.9 The development and implementation of policy at the European level is a dynamic and ongoing process. Playing a leading and influential role in the European process is a key aim for the Government. The discussions in which the UK, as a Member State of the European Union, is involved, are both broad and detailed. For example, as at March 1999, discussions on 20 separate waste related issues were ongoing:

- **the Community strategy for waste management** – the strategy is due to be reviewed shortly

- **regulations on waste management statistics** – the objective of this Regulation is to establish a framework for the production of Community statistics on waste production and management. The Regulation will require Member States to regularly provide data covering municipal waste and waste produced by a range of economic activities. Detailed information will be required on different waste categories and on recycling, incineration, composting and disposal of waste

- **communication on recycling** – the Commission has published a study of the competitiveness of the EU recycling industries. The study, in the form of a Commission Communication, was agreed by the Environment and Industry Councils, and included in a proposal to set up a Recycling Forum. The first session was held in January 1999, and participants were drawn from Member States, European trade associations and non-governmental organisations. The Forum reported back to the Commission early in 2000 with recommendations on how to best improve the competitiveness and performance of the recycling industry. The Commission is considering these recommendations

- **hazardous waste list** – proposed amendments the hazardous waste Directive applies to wastes on the Hazardous Waste List. The list consists of about 240 wastes. In 1997, the Commission began the process of reviewing proposals to amend the list. This process could result in the addition of up to a further 240 wastes to the list

- **prevention of waste** – waste reduction features in the 1996 Council Resolution on Waste, which reiterated its conviction that waste prevention should be the first priority for all waste policy in relation to minimisation of waste production and the hazardous properties of waste. The Commission recently indicated that a Communication is planned to promote the prevention of waste

- **proposed marking of packaging Directive** – the proposed Directive on marking of packaging and on a conformity assessment procedure was published by the Commission in 1996. It resulted from the 1994 EC Directive on packaging and packaging waste (94/62/EC)

- **draft end-of-life vehicles Directive** – this Directive aims to improve the environment by increasing the proportion of vehicles recovered through improved treatment of scrapped vehicles, ensuring that hazardous materials recovered from the vehicles do not go to landfill, and ensuring that manufacturers make vehicles more recyclable in the future

- forthcoming proposal for a **Directive on waste from electrical and electronic equipment** – the Commission has indicated its intention to propose a Directive on the management of waste from electrical and electronic equipment. The Commission has been considering the following proposals for this sector: phasing out (with some exceptions) the heavy metals lead, cadmium, mercury and hexavalent chromium, and certain halogenated flame retardants; specifying that producers should be responsible for setting up collections systems and achieving targets for collecting end-of-life equipment; establishing re-use and recycling targets and specifying that producers should be responsible for achieving these; and specifying how separately collected end of life equipment should be treated

- **household waste and the hazardous waste Directive** – the Hazardous Waste Directive requires the Commission to bring forward proposals for including household waste in the Directive. Such a proposal may require segregation of up to 20 specified hazardous household wastes. It may also ban the mixing of one category of household hazardous waste with another

- **directive on batteries and accumalators** containing certain dangerous substances – the Commission intend to revise the Batteries Directive. An early draft proposal includes provisions to: extend the scope of the Directive to cover all batteries, not just those containing lead, mercury or cadmium; set quantified collection and recycling targets and require design of some equipment to permit easy removal of spent batteries; and ban the use of nickel-cadmium batteries (except for certain specified uses)

- Commission proposal to amend the **Sewage Sludge in Agriculture Directive** – the Commission has indicated its intention to propose a revision of the current Directive (86/278/EEC) to update it and to take into account the increased quantities of sewage sludge being produced due to the implementation of the urban waste water treatment Directive. The broad intention of the review is to enhance the level of protection of the environment offered by the present Directive

- **Waste Oils Directive** – proposed amendments – the main provisions of the 1987 Directive on waste oils are aimed at securing collection and disposal of waste oil without causing harm to human health or the environment and in particular to give priority to regeneration over recovery (by combustion) and disposal. The Commission has indicated that it intends to review the Directive, to set targets for collection and recycling – including a target for the recycled content of new oils

- proposed **construction and demolition wastes Communication** – this is a significant waste stream in terms of volume and was the subject of a priority waste stream report in 1995. Since then, consultants have been undertaking a research project for the Commission, who have indicated they are planning a Communication on construction and demolition waste

- proposal for a **Composting Directive** – the Commission have indicated they plan to produce a draft proposal for a composting Directive, though the timescale has yet to be determined. This could include mandatory source seperation of biodegradable material, and limits on contaminants in composted material. It may also include targets for composting

- **proposed Communication on PVC** – the Commission has announced that it plans to produce a Communication on PVC waste

- **proposed Directive on marking of plastics** – the Commission intends to bring forward a proposal on plastic wastes, possibly to promote separation and recycling

- **healthcare wastes** – this waste stream was the subject of a priority waste stream report, which remains unpublished. The Commission has announced that it plans to produce a Communication on healthcare wastes

- **Directive on Packaging and Packaging Wastes** – the recovery and recycling targets in the Directive (94/62/EC) are to be reviewed, and new targets for 2002-2006 fixed by the Council, before the end of 2000

- proposal for a **Directive on the Incineration of Waste** – the proposed Directive aims to reduce emissions to air, water and land from the incineration of non-hazardous wastes, through the extension of emission controls to processes outside the scope of Directive 94/67/EC on the incineration of hazardous waste, and the imposition of more stringent controls on municipal waste incineration plant than those required by the two 1989 Directives (89/369/EEC and 89/429/EEC), which were concerned only with certain emissions to air

  The proposal involves uniform emission limits (to air and water) and controls over solid residues. It applies to a range of different types of plant: municipal waste, sewage sludge and clinical waste incinerators, plus a variety of less common and smaller plant, for example incinerators burning treated waste wood and waste oil, and co-incineration processes (including waste oils and tyres in cement kilns) and the combustion of wastes such as sewage sludge in conventional electricity generation plants

- **Waste Shipments Regulation** – the Waste Shipments Regulation establishes controls over transfrontier movements of waste within, into and out of the EU. It also implements the projected Basel Convention ban on exports of hazardous waste from OECD to non-OECD countries. Subsidiary Regulations govern exports of green list (non-hazardous) waste to non-OECD countries. The Commission, with Member States, is currently reviewing the Regulation

B.10 Discussions on these issues will not necessarily result in changes to the way waste is managed either in the UK or across Europe, though the fact that an issue is being discussed does reflect the desire of some – and sometimes all – Member States for common action to be taken in these areas.

Waste legislation in England and Wales

B.11 Over the past 25 years waste legislation in England and Wales has been adapted and improved to bring it into line with evolving European legislation on waste management. The twin aims of this legislation is to ensure:

- that waste is collected, managed, transported, stored, recovered and disposed without harm to human health or the environment

- that responsible authorities develop and adopt effective plans for managing and disposing of waste arising or imported into their localities.

Making Waste Work

B.12 In 1995, the Government introduced legislation establishing the Environment Agency as a body to regulate – among other things – the management and disposal of waste in England and Wales. This legislation (the 1995 Environment Act) also amended existing legislation (including the 1990 Environmental Protection Act) to rationalise the requirements to plan effectively for waste, including the preparation of national waste strategies.

B.13 In addition, in December 1995, the Government published a White Paper on waste management – *Making Waste Work*. That White Paper set out a number of strategies for improving the performance on waste of England and Wales. It also set out two primary (and a number of secondary) aspirational targets for waste management. The primary targets were:

- to reduce the proportion of controlled waste going to landfill from 70% to 60% by 2005

- to recover value from 40% of municipal waste by 2005.

B.14 The secondary indicative targets focussed on a number of key issues, including one of recycling or composting 25% of household waste by 2000.

B.15 The Government and the National Assembly for Wales support the main aims of *Making Waste Work* – though they believe we need to go much further towards waste reduction, recovery and recycling – and have used the principles underlying that White Paper to help develop this new waste strategy for England and Wales.

B.16 This strategy introduces a new set of targets and indicators, set out in Part 1 of this strategy. They reflect the view that for targets to be effective in changing behaviour towards a desired end, they have to be challenging but achievable, underpinned by a clear course of action, and capable of proper measurement and public explanation.

WASTE LEGISLATION

B.17   The following primary and secondary legislation all currently have an impact on waste management in England and Wales:

- Control of Pollution Act 1974
- Local Government Act 1985
- Control of Pollution (Amendment) Act 1989
- Environmental Protection Act 1990
- Town and Country Planning Act 1990
- Planning and Compensation Act 1991
- Environment Act 1995
- Finance Act 1996
- Merchant Shipping and Maritime Security Act 1997
- Town and Country Planning General Development Order 1988, SI 1813
- Controlled Waste (Registration of Carriers and Seizure of Vehicles) Regulations 1991, SI 1624
- Environmental Protection (Duty of Care) Regulations 1991, SI 2839
- Controlled Waste Regulations 1992, SI 588 (as amended)
- Environmental Protection (Waste Recycling Payments) Regulations 1992, SI 426
- Waste Management Licensing Regulations 1994, SI 1056 as amended
- Town and Country Planning (General Permitted Development) Order 1995, SI 418
- Town and Country Planning (General Development Procedure) Order 1995, SI 419
- Special Waste Regulations 1996, SI 972 (as amended)
- Chemicals (Hazard Information and Packaging for Supply) Regulations 1996
- Producer Responsibility Obligations (Packaging Waste) Regulations 1997, SI 648
- Packaging (Essential Requirements) Regulations 1998, SI 1165

B.18   As of 1 July 1999, the National Assembly has responsibility for making most secondary legislation in Wales, including that relating to waste management.

The legal definition of waste

B.19   The definition of waste in force in England and Wales is the definition given in Article 1(a) of the amended Framework Directive on Waste, which states that: *waste shall mean any substance or object in the categories set out in Annex I which the holder discards or intends or is required to discard.* There are currently 16 categories in annex 1 to the Directive.

- production or consumption residues not otherwise specified below

- off-specification products

- products whose date for appropriate use has expired

- materials spilled, lost or having undergone other mishap, including any material, equipment, etc contaminated as a result of that mishap

- materials contaminated or soiled as a result of planned actions (eg residues from cleaning operations, packaging materials, containers, etc)

- unusable parts (eg reject batteries, exhausted catalysts, etc)

- substances that no longer perform satisfactorily (eg contaminated acids, contaminated solvents, exhausted tempering salts, etc)

- residues of industrial processes (eg slags, still bottoms, etc)

- residues from pollution abatement processes (eg scrubber sludges, baghouse dusts, spent filters, etc)
- machining/finishing residues (eg lathe turnings, mill scales, etc)

- residues from raw materials extraction and processing (eg mining operations, oil field slops, etc)

- adulterated materials (eg oils contaminated with PCBs, etc)

- any materials, substances or products whose use has been banned by law

- products for which the holder has no further use (eg agricultural, household, office, commercial and shop discards, etc)

- contaminated materials, substances or products resulting from remedial action with respect to land

- any materials, substances or products which are not contained in the above categories

B.20   As can be seen, determining whether something is waste is not simple. The question of whether or not a substance or object is waste is one which must be determined on the facts of the case and in the light of judgements issued by the European Court of Justice (such as the *Mayer Parry* judgement). Since the adoption of the Directive the European Court of Justice has considered several cases on the definition of waste. Advice to help people decide whether something is waste can be found in DOE Circular 11/94.

ANNEX C

Waste Strategy 2000: Regulatory Impact Assessment

C.1 The purpose of this Regulatory Impact Assessment (RIA) is to assess the impact of Waste Strategy 2000, the waste strategy for England and Wales, identifying costs and benefits to business, charities, voluntary bodies and government.

C.2 As is the nature of a strategic document, the waste strategy does not contain specific regulatory proposals which could be subject to a detailed assessment of costs and benefits. Rather it draws together information on a wide variety of current and proposed regulatory and non-regulatory measures, most of which are already or will be the subject of individual RIAs or similar assessments. Some of these measures do not stem specifically from the waste strategy, but are in place in order to fulfil UK obligations under EC Directives and other policy objectives. The costs and benefits of these measures are therefore not attributable to the waste strategy. Other measures are not expected to impose a burden on organisations such as business, charities or voluntary bodies, and are therefore not discussed in this RIA.

C.3 Appendix A summarises the existing and proposed instruments discussed in the waste strategy which are already or will be subject to individual RIAs or which are not expected to impose a regulatory burden.

C.4 This RIA focuses on:

- the targets for municipal waste recovery, recycling and composting

- the target for recovery of commercial and industrial waste

- the choice of policy instrument to reduce the quantity of biodegradable municipal waste going to landfill.

PURPOSE AND INTENDED EFFECTS OF THE WASTE STRATEGY

C.5 The waste strategy has several purposes. These include:

- fulfilling our legal obligation under the 1990 Environmental Protection Act (as amended by the 1995 Environment Act) to prepare a national waste strategy. This requirement flows from our obligation under the Waste Framework Directive (75/442/EEC as amended by Council Directive 91/156/EEC) to develop a waste management plan

- helping to fulfil requirements under the Landfill Directive (99/31/EC), the Hazardous Waste Directive (91/689/EEC) and the Packaging Waste Directive (94/62/EC)

- ensuring that waste management contributes fully to the goal of more sustainable development

C.6   The effects of the waste strategy are intended to be:

- greater diversion of waste from landfill, and substantial increases in recycling and energy recovery

- engagement of the public in increased re-use and recycling of household waste

- to curb the growth in waste arisings

- for waste managers to increasingly base waste management decisions on an assessment of the best practicable environmental option

- to promote greater awareness of the end of life environmental impacts of a product, beginning at the product design stage

- the development of the market for secondary (recycled) materials

C.7   The purpose of **targets for municipal, commercial and industrial waste** management is to describe what the Government considers to be the appropriate balance between different waste management options. The targets indicate the scale of the challenge ahead in achieving our obligations under the various European legislation and domestic policy objectives listed in Appendix A and moving towards more sustainable waste management. The targets will help to reduce uncertainty for waste managers and waste management companies in planning for new facilities and services and in making investment decisions.

C.8   The purpose of **reducing the quantity of biodegradable municipal waste landfilled** is to contribute towards UK measures to meet targets set in Article 5 of the Landfill Directive.

OPTIONS

C.9   The options for setting **targets for recycling and recovery of municipal waste** include:

- do nothing

- setting targets that solely reflect the need to divert biodegradable municipal waste from landfill in order to comply with the Landfill Directive[1] – the 'bio' option

- setting goals that include diversion of non-biodegradable waste, which is not specifically required by the Directive – the 'bio and non-bio' option

Appendix C provides a summary of modelling work undertaken to assess the costs and benefits of these options. The 'do nothing' option is represented by the 'base case' scenario, in which most municipal waste continues to be landfilled. The 'bio' option is represented by the 'mix 1' scenario (in which all waste diversion required to meet the Landfill Directive targets is achieved through incineration) and the 'mix 2' scenario (where there is composting as well as inceneration). The 'non-bio' option is represented by the 'mix 3' scenairo (in which there are substantial increase in recycling and composting as well as some incineration) and the 'mix 4' scenario (which involves yet higher levels of recycling and composting).

[1] Article 5 of the Landfill Directive requires Member States to reduce the amount of biodegradable municipal waste going to landfill to 25%, 50% and 65% below 1995 arisings by 2006, 2009, and 2016 or 2010, 2013 and 2020 in the case of Member States eligible to use a four-year derogation.

C.10   The options for setting targets for the recovery of **commercial and industrial waste** include:

- setting targets that reflect in aggregate the expected effects of existing and proposed policy instruments as listed in Appendix A

- setting targets that include diversion of wastes that are not covered by these instruments

C.11   The options for **reducing the quantity of biodegradable municipal waste landfilled** are the issue of permits to restrict the quantity of biodegradable municipal waste that can be landfilled, either to local authorities or to landfill operators. The September 1999 consultation paper *Limiting Landfill* discussed other options, which were bans on landfilling biodegradable municipal waste or specific types of biodegradable municipal waste. Consultation responses came out strongly against these options, and *Limiting Landfill* commented that they were likely to be more costly than permits. They are not considered further in this RIA.

C.12   A range of permit design issues flow from the choice between landfill operator permits and local authority permits, for example whether to make permits tradable, how to allocate permits, and duration of permit life. These issues are not discussed in detail in this RIA, as such decisions will be the subject of a further consultation on the draft legislation required to introduce permits. The options for implementing other aspects of the directive will be the subject to a double consultation exercise: firstly a consultation in 2000 on the implementation of the regulatory controls on landfill in the Directive, followed by a second consultation paper on the draft legislation transposing these aspects into law.

BENEFITS

C.13   The benefits relevant to the measures proposed in the waste strategy are environmental. Most of the measures have a broad objective of curbing the growth in waste arisings, reducing the amount of waste that goes to landfill and increasing the amount of waste that is recycled, composted or has energy recovered from it.

C.14   Appendix B provides a summary of studies on the relative environmental benefits of different options for diverting waste from landfill. These studies suggest that there are environment benefits to be gained by a greater level of recovery, recycling and composting and that recycling certain materials can yield significant environmental benefits over energy recovery. These studies provide estimates of the economic value of environmental impacts of different waste management options. These values should be used with caution, due to the high level of uncertainty in quantifying and valuing impacts, particularly impacts on human health. However, they provide a broad indication of the relative environmental impacts of different waste management options.

C.15   The modelling work to assess the costs and benefits of options for **targets for the recovery and recycling of municipal waste** (see Appendix C for details) provides estimates of the level of waste recycling, composting and recovery that would be achieved using different mixes of waste management options (see Table C1). The targets are:

- to recover value from 40% of municipal waste by 2005

- to recover value from 45% of municipal waste by 2010

- to recover value from 67% of municipal waste by 2015

- to recycle or compost at least 25% of household waste by 2005

- to recycle or compost at least 30% of household waste by 2010

- to recycle or compost at least 33% of household waste by 2015

Table C1: **Results of modelling work for municipal waste**						
Mix	**Household waste recycling and composting**			**Municipal waste recovery**		
	2005	2010	2015	2005	2010	2015
Base case	7% – 9%	6% – 9%	5% – 9%	14% – 17%	12% – 17%	10% – 17%
Mix 1	7% – 9%	6% – 9%	5% – 9%	26% – 38%	34% – 54%	66% – 80%
Mix 2	15% – 19%	16% – 22%	15% – 22%	26% – 37%	29% – 49%	59% – 75%
Mix 3	23% – 30%	25% – 35%	25% – 35%	31% – 42%	35% – 52%	61% – 78%
Mix 4	24% – 32%	30% – 42%	34% – 47%	33% – 40%	39% – 51%	56% – 76%
Ranges are a result of differing assumptions about the growth in waste arisings (0–3%) and level of participation in kerbside recycling and composting schemes; see Appendix C for descriptions of base case and mixes 1–4.						

C.16   Mixes 3 and 4 are designed so that the levels of recycling, composting and energy recovery achieved are roughly consistent with the targets. Mixes 1 and 2 are not expected to lead to levels of recycling, composting and energy recovery sufficient to meet the targets. If the external cost estimates shown in Table C4, Appendix B were applied, mixes 1 and 2 would have £1.4 billion to £2.0 billion and £0.7 billion to £2.7 billion net environmental benefits (present value of benefits to 2020) over the case case. Mixes 3 and 4 would have £6.8 billion to £11.5 billion and £8.4 billion to £13.8 billion net environmental benefits (present value of benefits to 2020) over the base case. Present value of benefits to 2020 means the sum of discounted annual net benefits for the years 2001 to 2020. The discount rate used in 6%.

C.17   The **targets for the recovery and recycling of commercial and industrial waste** reflect in aggregate the expected effects of a range of existing and proposed policy instruments (included in list in Appendix A).

C.18   The benefits resulting from each of these instruments is or will be discussed in the RIAs of each instrument. These will include a reduction in the risks associated with landfill disposal from the banning of specified wastes such as tyres and liquid wastes, and a greater level of diversion to recycling and energy recovery.

C.19   It is estimated that these instruments will divert around 11 million tonnes from landfill by 2004, which is expected to meet the proposed target of reducing commercial and industrial waste going to landfill to 85% of 1998 levels by 2005 if annual growth in commercial and industrial waste arisings does not exceed 1.8% per year.

C.20   At present there is insufficient information to assess whether setting targets that go beyond this level would yield net benefits. A voluntary approach is therefore proposed in the interim, and it is assumed that industry will only set itself goals that yield net benefits.

C.21   Both options for **reducing the quantity of biodegradable municipal waste landfilled** allow some flexibility in how reductions are achieved: through reduction in arisings, or diversion from landfill through composting, recycling or incineration with energy recovery. Waste managers will be able to choose the best option for reducing the amount of biodegradable municipal waste sent to landfill, taking account of environmental impacts as well as financial costs.

C.22   Local authority permits could provide a better way for the environmental benefits of meeting the Landfill Directive targets to be maximised. There would be less uncertainty about the future availability of permits and local authority waste managers may be better able to plan to meet the targets, taking account of the Best Practicable Environmental Option for dealing with waste in their area.

C.23   The environmental performance of the instrument could be strengthened if trading of permits is allowed, as trading would provide greater flexibility in choosing the best way to deal with waste. Trading may also provide greater impetus to local authorities to cooperate with neighbouring authorities to procure waste management services, strengthening the capability of individual authorities to maximise the environmental benefits of meeting the Landfill Directive targets.

C.24   Waste management companies that provide biodegradable municipal waste reduction or landfill diversion services could benefit from introduction of either instrument.

BUSINESS SECTORS AFFECTED

C.25   The **targets for recycling and recovery of municipal waste** will have positive and negative impacts on the waste management industry. Local authorities will look to the private sector to provide services to divert waste from landfill, whilst demand for landfill services should decline over time. There may also be indirect effect on the virgin raw materials and energy sectors.

C.26   The **targets for commercial and industrial waste** as currently proposed will have an impact on those sectors affected by the instruments listed in Appendix A. However the waste strategy will not impose an additional burden, as the targets represent the aggregate level of diversion expected as a result of those instruments.

C.27   Landfill operators, waste hauliers and other waste management service providers would be affected by both options for **reducing the quantity of biodegradable municipal waste landfilled**. These may also be indirect effects on the virgin raw materials and energy sectors.

COMPLIANCE COSTS FOR BUSINESS, CHARITIES AND VOLUNTARY ORGANISATIONS

C.28   The **targets for recycling and recovery of municipal waste** will not impose direct compliance costs on business, charities and voluntary organisations. The **targets for commercial and industrial waste** as currently formulated will not impose compliance costs over and above the compliance costs identified in the RIAs for the instruments listed in Appendix A. Any voluntary approaches for particular wastes should allow businesses in the relevant sector to identify options to reduce, recycle, compost or recover energy from these wastes that provide net financial or environmental benefits. The compliance costs to business are therefore expected to be low.

C.29   The compliance costs of both options for **reducing the quantity of biodegradable municipal waste landfilled** can be broken down into two types of cost: administrative costs and the costs of reducing biodegradable municipal waste arisings or diverting biodegradable municipal waste from landfill. Although under either option there may be a need for a parallel audit trail, administrative costs will mainly be incurred by landfill operators under landfill operator permit option, and by local authorities under the local authority permit option. The costs of reducing biodegradable municipal waste arisings or diverting biodegradable municipal waste from landfill will mainly be incurred by local authorities.

C.30   Both options – landfill operator permits and local authority permits – will result in loss of business to landfill operators. In the case of landfill operator permits, this loss may be offset by increasing revenues on biodegradable municipal waste landfill services provided, if permits are allocated free of charge. Under this option the price of biodegradable municipal waste landfill services is expected to be higher than under the local authority permit option.

C.31   The costs to business of landfill operator permits will include the administrative costs of applying for, managing the use of and reporting compliance with biodegradable municipal waste landfill permits. Landfill operators that hold permits may also be required to pay charges to the Environment Agency to cover the costs of monitoring compliance. There are 416 landfill sites licensed to accept municipal waste in England and Wales, owned by 111 waste management companies. It would be difficult to find a way to allocate permits to landfill opeators free of charge, due to a lack of existing data on the amounts of biodegradable municipal waste accepted at each of these 416 landfill sites. It may be that the only viable way to distibute landfill operator permits would be through auctioning, which would increase compliance costs to landfill operators and land to higher landfill prices.

C.32   If permits are tradable, landfill operators will incur additional administrative costs when they trade permits, both in making the transaction and in registering changes to their permit allocation. However it is to be expected that landfill operators will only trade permits if trading yields net benefits. There may be trading options that involve low administrative costs, such as transferring permits within a company or trading through a recognised permit exchange. For example, fees for trading Packaging Waste Recovery Notes through OM Environment Exchange total £1 per tonne (PRN prices vary over time and according to material, but range from £4 to around £100 per tonne). The costs of registering changes to permit allocations are expected to be low, as resource requirements for a registry are unlikely to exceed £50,000 per year (based on a staff of one full-time post and one part-time post).

CONSULTATION WITH SMALL BUSINESS

C.33   A Litmus Test has not been conducted, as most of the measures discussed in the strategy are at an early stage of development or are subject to separate RIAs. Responses to the consultation paper *Limiting Landfill* assisted the Department in taking on board the views of landfill operators on options for **reducing the quantity of biodegradable municipal waste landfilled**.

OTHER COSTS

C.34   The modelling work to assess the costs and benefits of options for **targets for the recovery and recycling of municipal waste** (see Appendix C for details) provides estimates of the costs of meeting the Landfill Directive targets using different mixes of waste management options.

C.35   Mixes 1 and 2 achieve the Landfill Directive targets at lower cost (£1.8 billion to £6.2 billion present value), as most of the recovery is achieved through incineration with energy recovery. However, it may not be possible to build the number of incinerators needed under mixes 1 and 2 because of the difficulty typically experienced in finding suitable sites and gaining public support for incinerators. Mixes 3 and 4 are more expensive (£3.4 billion to £7.7 billion present value) because they include the recycling of non-biodegradables, the majority of which would have been landfilled in the base case. Mixes 3 and 4 reflect the levels of recycling, composting and recovery necessary to meet the waste strategy targets. If mixes 1 and 2 are not expected to be viable because of the high number of incinerators they would require (60 to 166 and 41 to 129), mixes 3 and 4 (which indicate a need for 33 to 112 and 21 to 89 incinerators respectively) may represent the most likely mix of waste management options that will be used to meet the landfill Directive targets. In which case the targets for recovery and recycling of municipal waste would not impose a net additional cost on local authorities.

C.36   Local authorities will incur compliance costs under both options for **reducing the quantity of biodegradable municipal waste landfilled**. The bulk of their costs will be the costs of reducing biodegradable municipal waste arisings or diverting biodegradable municipal waste from landfill. Appendix C presents the results of cost modelling work that estimates the costs of meeting the Landfill Directive targets using different mixes of waste management options.

C.37   Both options – local authority permits and landfill operator permits – allow some flexibility in how to reduce biodegradable municipal waste arisings or divert waste from landfill, allowing waste managers to choose the best option for them, taking into account financial costs as well as environmental effects. As waste management costs vary in different locations, the total cost of meeting the targets will be reduced if permits are tradable. Trading permits allows most diversion to take place in locations where diversion is cheapest. The extent to which these cost reductions will be realised will depend on the level of trading that would occur. It could be expected that, at least initially, the level of trading would be lower under the local authority permit option than under the landfill operator permit option. However, local authorities could face higher waste management costs under the landfill operator permit option, as landfill prices are expected to be higher under this option than under the local authority permit option.

C.38   The costs to local authorities of local authority permits will include the administrative costs of applying for, managing the use of and reporting compliance with biodegradable municipal waste landfill permits. There are 135 Waste Disposal Authorities in England and Wales. There do appear to be a number of options for decidng how to distribute permits to local authorities free of charge – by grandfathering (for example on the basis of each local authorities' waste arisings) or by benchmarking (for example on the basis of the numbers of households in each local authority). Individual local authorities may be better or worse off under different grandfathering or benchmarking options. However trading permits will help to equalise costs across authorities. The Environment Agency would incur costs of monitoring compliance with local authority permits.

C.39   If permits are tradable, local authorities will incur additional administrative costs when they trade permits, both in making the transaction and in registering changes to their permit allocation. However it is to be expected that local authorities will only trade permits if trading yields net benefits. There may be trading options that involve low administrative costs, such as aggregating permits within a group of local authorities or trading through a recognised permit exchange. For example, fees for trading Packaging Waste Recovery Notes through OM Environment Exchange total £1 per tonne (PRN prices vary over time and according to material, but range from £4 to around £100 per tonne). The costs of registering changes to permit allocations are expected to be low, as resource requirements for a registry are unlikely to exceed £30,000 per year (based on a staff of one full-time post).

RESULTS OF CONSULTATION

C.40   A consultation paper on the waste strategy *Less Waste: More Value* was published in June 1998. A report of the responses to the consultation was published at the same time as the draft waste strategy – *A Way With Waste* – in June 1999. Consultation responses to *Limiting Landfill* were strongly in favour of local authority permits as a means of **reducing the quantity of biodegradable municipal waste landfilled**.

ENFORCEMENT, SANCTIONS, MONITORING AND REVIEW

C.41   No specific sanctions have as yet been proposed for either of the options for reducing the quanity of biodegradable municipal waste landfilled. Enforcement of landfill opeator permits seems feasible, as monitoring and enforcement could be conducted as part of existing environmental regulation of landfill sites. Local authority permits would require new monitoring arrangements. Sanctions available will depend on the powers confered under enabling legisation.

SUMMARY AND RECOMMENDATIONS

C.42   The proposed **targets for the recovery and recycling of municipal waste** may not impose net additional costs if it is expected that a broad mix of waste management options will be required to meet the Landfill Directive targets; and that reliance on incineration and composting is unlikely to be viable. It is therefore recommended that the proposed targets are adopted.

C.43   The proposed **targets for the recovery and recycling of commercial and industrial waste** do not impose net additional costs or benefits. It is therefore recommended that the proposed targets are adopted.

C.44   Consultation responses strongly suggest that local authority permits are the preferred option for **reducing the quantity of biodegradable municipal waste landfilled**. A comparison of the two instruments shows that neither instrument presents clear advantages over the other. Local authority permits may provide a better option for maximising the environmental benefits of meeting the Landfill Directive targets; the total administrative costs could be lower because of the lower number of organisations requiring permits. Tradable landfill operator permits may provide a better option for minimising the financial costs of diverting waste in order to meet the targets as landfill operators may be more inclined to trade permits; however total administrative costs could be higher because of the higher number of organisations requiring permits.

C.45   Landfill operator permits could result in higher prices for landfilling biodegradable municipal waste, which would have the effect of increasing local authority waste management costs but which (if allocated free of charge) would allow landfill operators to offset losses due to the forced reductions in landfilling biodegradable municipal waste. It could be difficult to arrive at means of allocating permits to landfill operators free of charge, due to lack of data on biodegradable municipal waste accepted at individual landfill sites. Allocation to local authorities on the basis of numbers of households, population or waste arisings could provide a means of allocating permits free of charge to local authorities. Monitoring compliance with and enforcement of landfill operator permits could be added to existing environmental regulatory activities of the Environmental Agency. Local authority permits would require new arrangements for monitoring. Sanctions available will depend upon powers conferred by enabling legislation.

C.46   The advantages and disadvantages of the two options are therefore finely balanced. It is therefore recommended that local authority permits, the strong favourite in consultation responsed, be given further consideration.

     Contact point: Joanna Coulton, Environment Protection Economics Division, Department of the Environment, Transport and the Regions, Zone 5/E4 Ashdown House, 123 Victoria Street, London SW1E 6DE. Tel: 0171 890 6462; Fax: 0171 890 6419

APPENDIX A

Table C2: Main existing and proposed measures discussed in the waste strategy but excluded from this RIA

Instrument	Status	Most recent assessment of costs and benefits
Landfill Tax	Existing	Financial Statement and Budget Report 1999, 2000
Landfill tax credit scheme	Existing	Landfill Tax Regulations SI 1996 No. 1527, amended SI 1999 No. 3270
Landfill Directive	Common Position adopted; implementation mechanisms to be proposed	Draft RIA on Common Position text April 1999; a consultation paper and RIA on measures to implement the Article 5 targets published in September 1999; a consultation paper on implementation of other aspects of the Directive will be published in 2000
Packaging Regulations	Existing	RIA Packaging (Essential Requirements) Regulations May 1998 RIA Producer Responsibility Obligations (Packaging Waste) Regulations March 1997 RIA Producer Responsibility Obligations (Packaging Waste) Amendment May 1999 RIA Producer Responsibility Obligations (Packaging Waste) Amendment 2 December 1999
Batteries and Accumulators Directive	Proposed revision of existing Directive	RIA to be prepared
Directive on End of Life Vehicles	Agreement on Common Position has been reached	Draft RIA published September 1999
Directive on Waste Electrical and Electronic Goods	Draft proposal expected in summer 2000	RIA to be prepared
Further Producer Responsibility Initiatives	Proposed	The costs and benefits of any further producer responsibility initiatives will be considered when choosing the type and design of instrument
Extension of hazardous waste list	Proposed (implementation into UK law by 1/1/2002)	RIA to be prepared in 2001
Waste Incineration Directive	Draft	Regulatory and Environmental Impact Assessment of the Proposed Wate Incineration Directive Entec UK Ltd Jan 1999 Further work for the Regulatory and Environmental Impact Assessment of the Proposed Waste Incineration Directive Entec UK Ltd March 1999 Regulatory and Environmental Impact Assessment of Amending the Hazardous Waste Incineration Directive to align with the Proposed Waste Incineration Directive Entec UK Ltd March 1999 Regulatory and Environmental Impact Assessment of Potentially Tighter Standards and Requirements of the Proposed Waste Incineration Directive Entec UK Ltd June 1999
The Local Government Bill (Best Value)	Draft	Explanatory notes accompanying the draft Bill 25 March 1999

Table C2: Main existing and proposed measures discussed in the waste strategy but excluded from this RIA (continued)

Instrument	Status	Most recent assessment of costs and benefits
Publication of new PPG10 and PPG11 guidance	Revision of existing guidance	Revised PPG10 published September 1999 Revised PPG11expected May 2000
Integrated Pollution Prevention and Control (IPPC)	Existing	PPC act introduced 1999. RIA to be prepared for proposed PPC regulations
PPC permits for landfill sites	Proposed	To be included in the consultation paper on implementing the Landfill Directive, to be published in 2000.
Special Waste Regulations	Existing	Special Waste Regulations SI 1996 No. 972, amended SI 1996 No. 2019 and SI 1997 No. 251
Operator Pollution Risk Appraisal (OPRA)	Proposed	Was the subject of a consultation paper April 1999 'Waste Management Licensing Risk Assessment. Inspection Frequencies – Operator Pollution Risk Appraisal for Waste'. Amendments to Statutory Guidance expected in 2000
Derogation from excise duty on waste oils	Existing	Expires 2001. Renewal to be considered
Aggregates Tax	New	Financial Statement and Budget Report 2000

Measures not expected to impose a regulatory burden on businesses, charities or voluntary organisations:

- Waste and Resources Action Programme (proposed new body to promote sustainable waste management)
- Incentive schemes and local authority consideration of targets to curb growth in municipal waste arisings
- National Waste Awareness Initiative/Are You Doing Your Bit campaign
- Pilot projects for government departments to purchase recycled products
- Environmental Technology Best Practice Programme

APPENDIX B

ENVIRONMENTAL BENEFITS OF DIFFERENT RECOVERY METHODS

Cb.1   Although recycling activities have negative impacts on the environment (through transporting and reprocessing waste to produce secondary resources), recycling has net environmental benefits because:

- recycling waste means that less waste is sent to landfill and other disposal options which have greater pollution and disamenity effects than recycling

- for some materials, producing goods from secondary resources reduces the environmental impacts of production through reduced energy use and reduced pollution

- using more secondary resources means using less primary resources (renewable and non-renewable natural resources) and the management, extraction, processing and distribution of primary resources can have environmental impacts that are greater than the environmental impacts of obtaining secondary resources from waste

- current rates of depletion of non-renewable natural resources and methods of managing renewable natural resources may not be consistent with sustainable development

Cb.2   Diverting waste from landfill to recycling therefore substitutes a practice that has only environmental costs with a practice that has net environmental benefits.

Cb.3   Recovering energy from waste, mainly by incineration, can have net environmental benefits relative to landfilling waste because:

- transport distances for sending waste to an incinerator are typically less than those for landfill

- generating electricity from waste means that less electricity needs to be generated from fossil fuels; this provides an environmental benefit particularly where the waste incinerated contains non-fossil carbon and where the electricity source displaced is more polluting than waste incineration

- waste incinerators can also provide energy in the form of heat (Combined Heat and Power)

Cb.4   Several studies have been published which attempt to value the environmental costs and benefits of different waste management options. It is difficult to make comparisons between these studies, as they adopt different assumptions about technology used, emission factors, environmental impacts of emissions and economic values of environmental impacts. They do not all have the same coverage of environmental impacts and some emissions and impacts are omitted due to absent data. Some studies do not adequately report the uncertainty in their estimates of environmental costs, and these are likely to be significant.

Cb.5 Most of the work done so far looks at single waste management options (ETSU 1996), single materials (BNMA 1995), or compares the environmental costs of incineration and landfill (CSERGE 1993). Only one major study (Coopers & Lybrand 1997) compares incineration, landfill and recycling the different fractions of MSW. The results of these studies generally support the view that recycling has net environmental benefits over incineration and landfill (with the possible exception of certain materials, such as plastic film). The results of the Coopers & Lybrand (1997) study are shown below (Table C4, options with negative external cost have net environmental benefits). However, these estimates seem to omit the costs of leachate from landfills, all disamenity costs and some of the environmental costs of processing for recycling.

Table C4: **External costs and benefits of different waste management options**	
Waste management option	External cost estimate, £ per tonne of waste, 1999 prices
Landfill	3
Incineration (displacing electricity from coal-fired power stations)	−17
Incineration (displacing average-mix electricity generation)	10
Recycling – Ferrous metal – Non-ferrous metal – Glass – Paper – Plastic film – Rigid plastic – Textiles	−161 −297 −929 −196 −69 17 −48 −66
Source: adapted from Coopers & Lybrand (1997)	

Cb.6 CSERGE (1993) suggests that existing rural landfill sites without energy recovery may have net external costs of between £1.90 and £10.90 per tonne (1999 prices). These estimates exclude disamenity costs which the study suggests could be significant. The same study suggests that urban incinerators with energy recovery may have net external benefits of up to £15.20 per tonne, through to a net environmental cost of £4.10 per tonne (1999 prices). However, this does not reflect potential disamenity impacts and it assumes that the electricity generated displaced is that of a coal-fired power station.

Cb.7 These estimates of external costs should be viewed with caution, due to the level of uncertainty in quantifying and valuing impacts, particularly those on human health. However they do provide an indication of the relative impacts of different waste management options – whilst recognising that the external costs per tonne of waste may vary according to location and the quantity of wast sent to each option.

Cb.8 It is likely that composting will play some part in meeting the Article 5 targets for the diversion of biodegradable municipal waste from landfill. This can be achieved through encouraging households to compost their own waste and/or collecting the waste and composting it at a central facility. The net effects of composting have not been quantified, but composting is likely to have a net environmental benefit as it helps avoid the environmental costs associated with landfill and displaces the need to extract, process and distribute other materials used as soil conditioners such as peat. This helps avoid the environmental costs associated with peat extraction, including habitat loss in particular. Composting, particularly home composting, may also have an additional benefit in terms of reduced distances travelled. Emissions from home composting are unlikely to cause environmental harm as they are diffuse.

REFERENCES

British Newsprint Manufacturers Association (BNMA, 1995) *Recycle or Incinerate? The future for used newspapers: an independent evaluation*, Swindon, UK: BMNA (Tel 01793 886086, Fax 01793 886182)

Coopers & Lybrand, CSERGE and EFTEC (1997) *Cost-Benefit Analysis of the Different Solid Waste Management Systems: Objectives and Instruments for the Year 2000*, Luxembourg: Office for Official Publications of the European Communities

CSERGE, Warren Spring Laboratory and EFTEC (1993) *Externalities from Landfill and Incineration*, London: HMSO

ETSU and Electrowatt Engineering (1996) *Economic Valuation of the Draft Incineration Directive*, Luxembourg: Office for Official Publications of the European Communities ISBN 92 828 0083 0A.1

APPENDIX C

ESTIMATING THE COSTS OF THE TARGETS FOR RECOVERY AND RECYCLING OF MUNICIPAL WASTE

Cc.1 The modelling work summarised in this Appendix refines and updates similar modelling presented in the RIA of the draft waste strategy *A Way With Waste*. Five mixes of waste management options were modelled to produce cost estimates over the period 2000 to 2020. A 'base case' was modelled so that the additional costs of meeting the Landfill Directive targets for biodegradable municipal waste and the waste strategy goals could be calculated. Mixes 1-4 were all designed to meet the Landfill Directive targets. Mixes 1 and 2 were expected to offer the least expensive way to meet the Landfill Directive targets, but were not designed to meet the waste strategy targets. Mixes 3 and 4 were designed to meet the waste strategy targets, which go further than the requirement of the Landfill Directive targets (the Landfill Directive targets do not require recycling of non-biodegradable waste).

Cc.2 Mixes 1-4 only represent a very generalised description of the possible mixes for meeting the Landfill Directive targets. Therefore, the modelling only provides a broad indication of the scale of costs involved. The mixes waste management options modelled were as follows:

- **Base Case** assumes current levels (in absolute terms) of recycling, composting, and energy recovery, with all other waste going to landfill.

- **Mix 1** continues current levels (in absolute terms) of recycling and composting; all waste diversion required to achieve the Landfill Directive targets is through incineration with energy recovery.

- **Mix 2** continues current levels (in absolute terms) of non-biodegradables recycling but increases levels of composting and paper recycling through kerbside collection of putrescible waste and 50% of waste paper; all other diversion required to achieve the Landfill Directive targets is through incineration with energy recovery.

- **Mix 3** increases levels of composting and recycling through kerbside collection of putrescible waste and dry recyclables; all other diversion required to achieve the Landfill Directive targets is through incineration with energy recovery.

- **Mix 4** is essentially the same as mix 3, but assumes a greater level of provision of kerbside collection services for putrescible waste and dry recyclables.

ASSUMPTIONS

Cc.3 The starting point for the modelling was waste arisings in England and Wales in 1995 (the base year for the landfill directive targets) and pattern of the arisings and disposals in England and Wales in 1998/9. It is assumed that England and Wales make use of the 4-year derogations for meeting the Landfill Directive targets.

Cc.4   A series of assumptions are made about the composition, biodegradability and recyclability of municipal waste, and the reject rate of waste collected for recycling (see Table C5). It is assumed that average the capacity of a Materials Recovery Facility (MRF) is 40 thousand tonnes per year; a composting station is 30 thousand tonnes per year; and an incinerator is 250 thousand tonnes per year. It is assumed that there will be sufficient demand for recyclables, compost and energy services from incinerators.

Table C5: **Assumptions used in modelling**				
	Waste composition	Biodegradable	Recyclable	Reject rate
Paper/card	32%	100%	65%	5%
Putrescible	21%	100%	90%	5%
Textiles	2%	50%	95%	5%
Fines	7%	60%	0%	0%
Misc. combustible	8%	50%	0%	0%
Misc. non-combustible	2%	0%	0%	0%
Ferrous metal	6%	0%	95%	5%
Non-ferrous metal	2%	0%	95%	5%
Glass	9%	0%	90%	5%
Plastic dense	6%	0%	33%	5%
Plastic film	5%	0%	0%	5%

Cc.5   In mixes 2 and 3, the level of provision of kerbside collection services for recycling and/or composting are ramped up to 60% of households in 2020; in mix 4, to 80% of households in 2020. All mixes were modelled at three rates of growth in waste arisings: 0%, 1% and 3% per year; and two participation rates: 75% and 55%. These participation rates are equivalent to the net result of a 'recognition rate' of around 95% (householders that recycle are proficient at recognising waste materials that can be recycled) and 'put out' rates of 80% and 60% (meaning 80% of household provided with the service actually use it). The variation in 'put out' rates has an effect on the costs of providing a kerbside collection service. The fewer households that tend to put out bins for emptying, the high the cost per tonne of providing the service – because the same kit and service has to be provided to collect a smaller amount of waste.

Cc.6   Different unit costs were therefore used to model the scenarios with different participation rates. The unit costs were estimated by the consulting firm Enviros Aspinwall and are shown in Table C6 below. These are the total gross costs; they do not include landfill tax and may not reflect the actual level of profit earned on services provided by the private sector or the level of revenue received from the sale of materials. The costs include collection, transfer and transportation to the disposal or recovery facility, as well as gate fees. They include operating and capital costs (which are annualised for conversion to costs per tonne). The unit costs for composting appear particularly high because of the high collection cost for compostables resulting from the assumption that the putrescible content of municipal waste is around 21%. In the high participation rate scenario, the unit cost for recycling is based on a 40 thousand tonnes per year materials recycling facility being used at a rate of two shifts per day, effectively doubling the capacity of the materials recycling facility to 80 thousand tonnes per year. The unit costs used in the modelling are indicative costs only: actual costs will depend on local factors.

Cc.7 The model produces estimates of total waste management costs for each year from 2000 to 2020. A discount rate of 6% is used to convert annual costs in future years into present value costs. The total cost of each scenario is then expressed as the sum of present value costs.

UNIT COSTS USED IN MODELLING

Table C6a: Unit costs used in modelling

	75% participation rate			55% participation rate		
	Urban	Suburban	Rural	Urban	Suburban	Rural
Collection						
Recycling	61	79	105	66	84	111
Composting	87	119	178	97	131	198
Landfill	15	21	33	15	21	33
Incineration	15	21	33	15	21	33
Treatment						
Recycling (40K)	23	23	23	23	23	23
Recycling (80K)	17	17	17	17	17	17
Composting	10	10	10	10	10	10
Landfill	25	25	15	25	25	15
Incineration	48	48	48	48	48	48
Gross costs						
Recycling (40K)	84	102	128	89	107	134
Recycling (80K)	78	96	122	83	101	128
Composting	97	129	188	107	141	208
Landfill	40	46	48	40	46	48
Incineration	63	69	81	63	69	81

Table C6b: Unit costs used in modelling

	75% participation rate			55% participation rate		
	Average urban, suburban & rural	Average urban & suburban	Costs used in modelling	Average urban, suburban & rural	Average urban & suburban	Costs used in modelling
Recycling (40K)		93			98	98
Recycling (80K)		87	87		92	
Composting		113	113		124	124
Landfill	45		45	45		45
Incineration	71		71	71		71

RESULTS

Cc.8 The estimated costs for the base case and mixes 1 – 4 are shown below (Table C7). The model also shows that increases in municipal waste arisings can increase waste management costs significantly; a 3% annual growth rate would increase municipal waste arisings from around 27.7 million tonnes (1998/9 arisings) to 53.1 million tonnes per year by 2020. Even in the base case, this would result in an increase in waste management costs of around £5.6 billion (present value of costs from 2000 to 2020).

Cc.9 The additional cost of mixes 1 and 2 over the base case are £1.8 billion to £4.9 billion and £2.7 billion to £6.2 billion respectively (present value of costs from 2000 to 2020). However, it may not be possible to find sufficient sites build the number of incinerators required under mixes 1 and 2. An indication of the numbers of different types of facility that could be required is given below (Table C8). The additional costs of mixes 3 and 4 over the base case are higher: £3.4 billion to £7.1 billion and £3.9 billion to £7.7 billion respectively (present value of costs from 2000 to 2020).

Table C7:	**Estimated costs of different mixes of waste management options from 2000 to 2020, £ billion**				
Mix	Growth rate/yr	Present value of costs from 2000 to 2020		Cost over base case	
Participation		75%	55%	75%	55%
Base case	0%	17.3	17.6	–	–
	3%	22.9	23.2	–	–
Mix 1	0%	19.1	19.4	1.8	1.8
	3%	28.8	28.1	4.9	4.9
Mix 2	0%	20.1	20.3	2.8	2.7
	3%	29.1	29.4	6.2	6.2
Mix 3	0%	20.8	21.0	3.5	3.4
	3%	29.2	30.3	7.0	7.1
Mix 4	0%	21.3	21.5	4.0	3.9
	3%	30.4	30.9	7.5	7.7

Estimated costs of different mixes of waste management options, 0-3% pa growth in municipal waste arisings

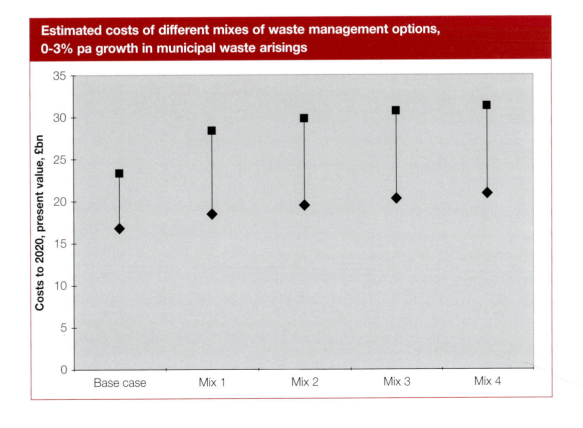

Table C8: **Estimated numbers of new waste management facilities that could be required for diversion of waste from landfill**				
Mix	Growth rate municipal waste/yr	Material recycling facilities	Composting stations	Incinerators
Base case	0%	0	0	0
	3%	0	0	0
Mix 1	0%	0	0	60
	3%	0	0	166
Mix 2	0%	0	99	41
	3%	0	196	128
Mix 3	0%	113	59	33
	3%	223	116	112
Mix 4	0%	160	84	21
	3%	316	164	89

Note: Numbers of material recycling facilities required could be halved if they are able to run two shifts per day increasing capacity to 80 thousand tonnes per year. The base case could require new landfill sites; mixes 3 and 4 could depend upon the development of new reprocessing facilities.

ANNEX D

Glossary of terms

Aggregates – sand and gravel and crushed rock used by the construction industry

Anaerobic digestion – a process where biodegradable material is encouraged to break down in the absence of oxygen. Material is placed into an enclosed vessel and in controlled conditions the waste breaks down into *digestate* and *biogas* (Part 2 section 5.75)

Basel Convention – the 1989 United Nations Basel Convention on the control of transboundary movements of hazardous wastes and their disposal provides a framework for a global system of controls on international movements of hazardous and certain other wastes (Part 2 Chapter 2 boxout – *exports and imports of waste*)

Best Practicable Environmental Option (BPEO) – a BPEO is the outcome of a systematic and consultative decision-making procedure which emphasises the protection and conservation of the environment across land, air and water. The BPEO procedure establishes, for a given set of objectives, the option that provides the most benefits or the least damage to the environment as a whole, at acceptable cost, in the long term as well as in the short term (Part 3 section 3.3)

Best Value – places a duty on local authorities to deliver services (including waste collection and waste disposal management) to clear standards – covering both cost and quality – by the most effective, economic and efficient means available (for waste collection – Part 2 section 4.99, for waste disposal – Part 2 section 4.106)

Central composting – large-scale schemes which handle kitchen and garden waste from households and which may also accept suitable waste from parks and gardens (Part 2 section 5.35)

Civic amenity waste – a sub-group of household waste, normally delivered by the public direct to sites provided by the local authority. Consists generally of bulky items such as beds, cookers and garden waste as well as recyclables

Clinical waste – waste arising from medical, nursing, dental, veterinary, pharmaceutical or similar practices, which may present risks of infection (Part 2 section 8.30)

Combined Heat and Power – a highly fuel efficient technology which produces electricity and heat from a single facility.

Commercial waste – waste arising from premises which are used wholly or mainly for trade, business, sport, recreation or entertainment, excluding municipal and industrial waste (Part 2 section 4.6, statistics included in Part 2 section 2.9)

Community sector – including charities, campaign organisations and not-for-profit companies.

Composting – an aerobic, biological process in which organic wastes, such as garden and kitchen waste are converted into a stable granular material which can be applied to land to improve soil structure and enrich the nutrient content of the soil (Part 2 section 5.32)

Construction and demolition waste – arises from the construction, repair, maintenance and demolition of buildings and structures. It mostly includes brick, concrete, hardcore, subsoil and topsoil, but it can also contain quantities of timber, metal, plastics and (occasionally) special (hazardous) waste materials (Part 2 section 8.43)

Controlled waste – comprised of household, industrial, commercial and clinical waste which require a waste management licence for treatment, transfer or disposal. The main exempted categories comprise mine, quarry and farm wastes. Radioactive and explosive wastes are controlled by other legislation and procedures

Duty of Care – applies to anyone who imports, produces, carries, keeps, treats or disposes of waste. Everyone subject to the duty of care has a legal obligation to comply with it and there are severe penalties for failing to do so. The Duty of Care does not apply to waste collection from households (Part 2 section 3.44)

EC Directive – a European Community legal instruction, which is binding on all Member States, but must be implemented through the legislation of national governments within a prescribed timescale

Energy recovery from waste – includes a number of established and emerging technologies, though most energy recovery is through incineration technologies. Many wastes are combustible, with relatively high calorific values – this energy can be recovered through (for instance) incineration with electricity generation (Part 2 section 5.53)

Environment Agency – established in April 1996, combining the functions of former local waste regulation authorities, the National Rivers Authority and Her Majesty's Inspectorate of Pollution. Intended to promote a more integrated approach to waste management and consistency in waste regulation. The Agency also conducts national surveys of waste arisings and waste facilities (Part 2 section 4.77)

Environmental Technology Best Practice Programme (ETBPP) – aims to demonstrate the benefits of managing resource use and reducing environmental impact to companies across the whole of the UK (Part 2 section 4.12)

Hazardous waste – see special waste

Home composting – compost can be made at home using a traditional compost heap, a purpose designed container, or a wormery (Part 2 section 5.48)

Household waste – this includes waste from household collection rounds, waste from services such as street sweepings, bulky waste collection, litter collection, hazardous household waste collection and separate garden waste collection, waste from civic amenity sites and wastes separately collected for recycling or composting through bring or drop-off schemes, kerbside schemes and at civic amenity sites (Part 2 section 4.41, statistics included in Part 2 section 2.14)

Incineration – is the controlled burning of waste, either to reduce its volume, or its toxicity. Energy recovery from incineration can be made by utilising the calorific value of paper, plastic, etc to produce heat or power. Current flue-gas emission standards are very high. Ash residues still tend to be disposed of to landfill (in general – Part 2 section 5.57, for special wastes – Part 2 section 6.23)

Industrial waste – waste from any factory and from any premises occupied by an industry (excluding mines and quarries) (Part 2 section 4.6, statistics included in Part 2 section 2.9)

Inert waste – waste which, when deposited into a waste disposal site, does not undergo any significant physical, chemical or biological transformations and which complies with the criteria set out in Annex III of the EC Directive on the Landfill of Waste

Integrated waste management – involves a number of key elements, including: recognising each step in the waste management process as part of a whole; involving all key players in the decision-making process; and utilising a mixture of waste management options within the locally determined sustainable waste management system (Part 2 section 3.1)

Integrated Planning Pollution and Control (IPPC) – is designed to prevent or, where that is not possible, to reduce pollution from a range of industrial and other installations, including some waste management facilities, by means of integrated permitting processes based on the application of *best available techniques* (Part 2 section 3.46)

Kerbside collection – any regular collection of recyclables from premises, including collections from from commercial or industrial premises as well as from households. Excludes collection services delivered on demand

Land use planning – the Town and Country Planning system regulates the development and use of land in the public interest, and has an important role to play in achieving sustainable waste management (Part 2 section 3.21)

Landfill sites – are areas of land in which waste is deposited. Landfill sites are often located in disused quarries or mines. In areas where there are limited, or no ready-made voids, the practice of *landraising* is sometimes carried out, where some or all of the waste is deposited above ground, and the landscape is contoured (Part 2 section 5.91)

Landspreading – is the spreading of certain types of waste onto agricultural land for soil conditioning purposes. Sewage sludge and wastes from the food, brewery and paper pulp industries can be used for this purpose (Part 2 section 5.86)

Licensed site – a waste disposal or treatment facility which is licensed under the Environmental Protection Act for that function (Part 2 section 3.35)

Life cycle assessment – can provide a basis for making strategic decisions on the ways in which particular wastesin a given set of circumstances can be most effectively managed, in line with the principles of Best Practicable Environmental Option, the waste hierarchy and the proximity principle (Part 2 section 3.12)

Minimisation – see reduction

Municipal waste – this includes household waste and any other wastes collected by a Waste Collection Authority, or its agents, such as municipal parks and gardens waste, beach cleansing waste, commercial or industrial waste, and waste resulting from the clearance of fly-tipped materials (Part 2 section 4.41, statistics included in Part 2 section 2.14)

Planning Policy Guidance Notes (PPGs) and Mineral Planning Guidance Notes (MPGs) – Government Policy Statements on a variety of planning issues, including waste planning issues, to be taken as material considerations, where relevant, in deciding planning applications

Producer responsibility – is about producers and others involved in the distribution and sale of goods taking greater responsibility for those goods at the end of the products life (generally – Part 2 section 4.35, for packaging waste – Part 2 Chapter 7)

Proximity principle – suggests that waste should generally be disposed of as near to its place of production as possible

Recycling – involves the reprocessing of wastes, either into the same product or a different one. Many non-hazardous industrial wastes such as paper, glass, cardboard, plastics and scrap metals can be recycled. Special wastes such as solvents can also be recycled by specialist companies, or by in-house equipment (generally – Part 2 section 5.19, for special wastes – Part 2 section 6.21)

Reduction – achieving as much waste reduction as possible is a priority action. Reduction can be accomplished within a manufacturing process involving the review of production processes to optimise utilisation of raw (and secondary) materials and recirculation processes. It can be cost effective, both in terms of lower disposal costs, reduced demand for raw materials and energy costs. It can be carried out by householders through actions such as home composting, re-using products and buying goods with reduced packaging (generally – Part 2 section 5.4, for special wastes – Part 2 section 6.17)

Re-use – can be practiced by the commercial sector with the use of products designed to be used a number of times, such as re-usable packaging. Householders can purchase products that use refillable containers, or re-use plastic bags. The processes contribute to sustainable development and can save raw materials, energy and transport costs (generally – Part 2 section 5.9, for special wastes – Part 2 section 6.21)

Self-sufficiency – dealing with wastes within the region or country where they arise

Separate collection – kerbside schemes where materials for recycling are collected either by a different vehicle or at a different time to the ordinary household waste collection

Special waste – is defined by the Control of Pollution (Special Wastes) Regulations 1980 as any controlled waste that contains any of the substances listed in Schedule 1 to the Regulations, or is dangerous to life, or has a combustion flashpoint of 21°C or less, or is a medical product as defined by the Medicines Act 1968 (Part 2 Chapter 6)

Sustainable development – development which is sustainable is that which can meet the needs of the present without compromising the ability of future generations to meet their own needs

Sustainable waste management – means using material resources efficiently, to cut down on the amount of waste we produce. And where waste is generated, dealing with it in a way that actively contributes to the economic, social and environmental goals of sustainable development

Treatment – involves the chemical or biological processing of certain types of waste for the purposes of rendering them harmless, reducing volumes before landfilling, or recycling certain wastes

Unitary Authority – a local authority which has the responsibilities of both Waste Collection and Waste Disposal Authorities

Waste – is the wide ranging term encompassing most unwanted materials and is defined by the Environmental Protection Act 1990. Waste includes any scrap material, effluent or unwanted surplus substance or article which requires to be disposed of because it is broken, worn out, contaminated or otherwise spoiled. Explosives and radioactive wastes are excluded

Waste arisings – the amount of waste generated in a given locality over a given period of time

Waste Collection Authority – a local authority charged with the collection of waste from each household in its area on a regular basis. Can also collect, if requested, commercial and industrial wastes from the private sector (Part 2 section 4.95)

Waste Disposal Authority – a local authority charged with providing disposal sites to which it directs the Waste Collection Authorities for the disposal of their controlled waste, and with providing civic amenity facilities (Part 2 section 4.104)

Waste hierarchy – suggests that: the most effective environmental solution may often be to reduce the amount of waste generated – *reduction*; where further reduction is not practicable, products and materials can sometimes be used again, either for the same or a different purpose – *re-use*; failing that, value should be recovered from waste, through *recycling, composting* or *energy recovery from waste*; only if none of the above offer an appropriate solution should waste be *disposed*

Waste management industry – the businesses (and not-for-profit organisations) involved in the collection, management and disposal of waste (Part 2 section 4.85)

Waste management licencing – licences are required by anyone who proposes to deposit, recover or dispose of waste. The licencing system is separate from, but complementary to, the land use planning system. The purpose of a licence and the conditions attached to it is to ensure that the waste operation which it authorises is carried out in a way which protects the environment and human health (Part 2 section 3.35)

Waste transfer station – a site to which waste is delivered for sorting prior to transfer to another place for recycling, treatment or disposal

Printed in the UK for The Stationery Office Limited on behalf of the
Controller of Her Majesty's Stationery Office
Dd 5067480 5/00 48003 Job No TJ0001663